# Essays In Exploring The Intrigues of Mankind In Enlightening His Mind And Soul.

*By*

**Harold J. Eisenberg**

## Dedication

I dedicate this book to my wife, my entire family, and to Carol, who made the book possible!

# About The Author

I studied at USC for six years to obtain my Doctor of Pharmacy degree and followed at the same university to get a BA degree in Zoology. I was offered to go to the Arctic with others to study the various life indigenous to the region and get a PhD, but I decided not to go. After completing my studies, I served in the United States Air Force as a veteran of the 616th Aircort, gaining invaluable experiences that shaped my perspective on discipline and service.

Following my military service, I went to Los Angeles State College and got a Master's in Psychology. I was then chosen to pursue a PhD at UCLA's Brain Institute, where I studied sense modalities and retroactive memory sensations abroad. I have practiced working with various mental disorders, including working with the American Indian population in treating alcoholic disorders, and at Patton State Hospital in San Bernardino with various mental health conditions. Additionally, I worked at Clinica Sierra Vista as a clinical pharmacist for nine years and also contributed to their psychology department.

I am also a student of ancient history, having visited Rome, Egypt, and Greece. My life has been dedicated to continuous learning and helping others through the knowledge I've acquired along the way.

# Table of Contents

# To Life Good Or Sad

Believing in yourself is one of the greatest aspects of a sound mind.

Smiling when feeling devilish, creating tensions regular to the soul and reason to resolve.

A learned mind practices wisdom, evolving through various problems in a lifetime.

Generations come and soon pass, caused by the sins of time.

Ordinary people walk the streets of cities, and only a few create changes in new designs.

Money makes the world go round and round, courage to spend on luxuries I could buy.

The pleasure of boarding a plane, flying above the streets below.

Cemeteries grow, and memories fade away, who will be on the death list for sorrow?

Gladness or sadness will always be part of those who live beneath the sun.

The strongest seek gladness, the weak endure sadness.

Man's future holds promises, which will not disappoint,

But those who plan not will be lost in the past of dreams.

From sunrise to sunset, another day has passed.

Does anyone know the last?

Man will be lost in memories of what they were all about.

Let not sorrow beckon you to its sorrowful way,

Enjoy the goodness that life shall bring on a sunny day!

Life can be a funny thing, bringing memories of joyful friends gathering and having fun.

Be of good humor and frolic on the day when the days left are but a few.

For in life, we can rejoice in living a wonderful life! Amen!

# Once In A While

Once in a while, life shall render a smile; other times, it renders a frown.

Times when high energy shall abound; other times, when you just lay down.

Once in a while, moments of great joy shall befall; other times, sadness shall install.

Once in a while, you shall find someone to adore; other times, loneliness shall implore.

Times when you really want to sing; other times, life is just a bad dream.

Once in a while, seeing the golden light in the day; other times, living in the darkness of dismay.

Once in a while, once in a while, will happen in the trials and tribulations that are a part of life.

# A Time Of Sadness And Gladness

God's Hebrew name is Yahuoh!

Homage and love unto Thee.

Being glad or sad in life transverses at any time.

You're sad about an outcome of life that you are certain will come,

And when it turns out great, you suddenly become glad.

When you're sad and become glad, it's the greatest feeling of them all.

Maybe the worst in thoughts become the best.

Is one of the greatest enlightenments to the peace of mind that does follow.

I had little faith that Yahuoh was blessing me in many ways and then delivered me from fear,

My faith suddenly does grow.

I was sad when someone died and glad to see the rising sun.

Gladness is great but even greater when preceded by sadness after all.

Trials and tribulations create times of sadness and fear,

But when Yahuoh delivers you from them, gladness and peace do befall.

Sadness can bring a tear, but gladness replaces it with a smile.

Sitting down in a moment of sadness, and a moment of gladness, jumping all around.

Emotions play an important part in life after all,

But a greater faith in Yahuoh is the best to befall.

# Reflections On Life

I shall find the wind upon it moving the branch.

Nobody is alone when accompanied by a good friend.

Nobody can gain respect until he performs a good deed.

Nobody can find peace until contentment lies within.

Nobody can really enjoy wealth until it is shared with the poor.

Let's enjoy the good fruits of life as time goes by.

# The Ephemeral Nature Of Life

Nowhere to be found, what was around yesterday—a flower picked from the garden is gone.

A butterfly flying yesterday has flown somewhere else; a tree losing a branch cannot be replaced.

A bird flying south to avoid the cold winter is gone until springtime—nothing stays the same.

The leaves fall from the tree, leaving it barren to see; tomorrow changes yesterday.

The squirrel isn't upon the branch when residing inside the trunk with stored nuts.

Beautiful things of today are stolen by tomorrow.

The sunshine becomes a little colder, changing the raindrops to snowflakes—nothing remains the same.

The blooming of the flower and the green grass are gone beneath the gray clouds in a brief time.

A couple of gray hairs and some wrinkles on my face—nothing stays the same.

# The Delusion Of The Perfect Life

Just a little bit of luck in guessing what tomorrow will bring.

If you guess but one thing, maybe fame and riches will transpire for you—you shall never know until tomorrow.

Imagine whatever you want in finding a clue in what life is all about.

Mamas and Papas fade away, and you stand on your own two feet to face whatever life shall bring.

Are you ready in manners and education to face the enigmas life shall present to you?

Stand tall, for here come the events of life—what are the proper outcomes to beat them all?

You must find a reason to be living, enjoying the simplest things to befall.

You need the armor of education and goodwill to succeed in what you attempt to do.

Know the current events in evaluating what's happening around you.

Pleasing a wife and knowing how to raise children properly is best for you.

Responsibility is discovered for future generations to secure a better life in peace and liberty for all.

# Life: A Dream Or Reality?

Who taught me the right things to do and bought my first ice cream cone? My father did those things. He's gone and passed away, not to be seen anymore. Is life just a dream?

Who kissed and held me in their arms? My mother did those things. Now gone and passed away. Is life just a dream?

Who gave me their love and my first bike? My grandparents did, after all. Now gone and passed away, not to be found. Is life just a dream?

Who played with me in times of swinging and playing ball? Cousins and friends with smiling faces. Now, many have passed away or are too old to play.

Is life just a dream?

Who taught me in school to read and that 7 plus 7 equals 14? My teachers did. Now gone and passed away. Is life just a dream?

Sisters who shared my life with their love—those still living far away.

Is life just a dream?

Aunts and uncles are sharing a turkey on Thanksgiving, and all their faces are fading away and gone.

Is life just a dream?

Sons and grandchildren are lost in distant places—only their voices on cell phones can be heard once in a while. Is life just a dream?

Hearing God's voice to serve Him in the name of Jesus Christ replaces all things lost in life forevermore. That's not a dream.

# The Storm Is Coming

The clouds grow darker in the sky,

Rumbling thunder loud sounds from far away.

Lightning flashes across the heavens, powerful to be seen.

Thunder sounding, lightning flashing, nature in command.

From up high, the raindrops begin to fall,

Falling faster as they hit the ground.

The wind begins to blow with its invisible hand,

Moving objects flying around everywhere.

The leaves move on the waving branches,

With the raindrops reaching the roots below.

Running to hide from the rain pouring down,

Not to get wet, those abiding in their homes witness

The powers of a storm.

The last flash and sound as the raindrops stop,

Landing on the grass and trees below.

The wind subsides as the sunlight puts a blue tint into the heavens above.

At last, at last, the rainbow does appear for all to see, a sign given by God.

Lying on my bed with a little tear,

When will the next storm appear?

# What Could Be

Lost in the desert with little water left to drink,

Concerned or maybe afraid, where can I find any water in this endless sand?

Suddenly, I see tropical trees surrounding the water and a pond below with birds flying to and fro.

Running and streaming water suddenly disappears as a mirage that I did see.

Shall I die of thirst and hallucinate like the mirage beneath the sizzling sun?

Many did not survive, their bones not found, hidden below the hot sand.

Faltering in the heat, getting weak, no mercy lies within the endless sand.

Suddenly, I heard someone saying, “Scene one is done, let’s take a break, another end.”

Two camels come into sight with humps to show, what a wonderful sight.

A miracle, after all.

Fiction or fact?

Past or present, survival in the desert is a trial.

# Education Yields Good Reason

Words come and shall go,

The greater the meaning, the more you shall know.

The wise may know about the elements combining to form the gases creating the atmosphere that supports all in life.

Many animals and plants abound in their many habitats.

How did the vast mountains begin and the endless seas fill?

How does a tornado work as it rapidly turns into a supercell, the strongest type of thunderstorm?

Will man control nature's storms that suddenly appear, destroying our cities and causing floods to renew?

Science, with research, discovers new concepts and pushes the boundaries of it all.

In the sea, an ocean of many unknowns to dogs and bees persists as a continent we must learn to hear.

Man is a carbon-based, living organic being,

Breathing oxygen and exhaling carbon dioxide beneath the sun.

If 10% use their brain, the nine left should define the reason for their blank mind.

A black hole and supernova are not fully known, and the trillion galaxies to be found.

What harmless viruses are newly found, but only one virus kills millions in another pandemic in its lethal way.

Silver, gold, and rare metals are wanted by almost everyone.

A boy with his soldier and a girl with her dolly evoke something meaningful in reality.

When the people wish for a good government, stand on the beach and in the city for it to befall,

Pay the entertainer to have a chance to have fun.

Let the children read and write,

Learn math and literacy.

Let wisdom prevail for all,

Education yields good reason for all.

# Good Vs Evil

A man knowing history can seek a better future for all,

And the present shall acknowledge the wisdom he speaks.

Those who entreat in useless values to destroy others shall go straight to evil's overthrow.

Righteousness shall prevail over this evil intent.

One nation invests its time in seeking freedom and liberty,

Another is the payment and bondage of its people.

A constant war of nations in good versus evil, one causing mountains to rise and another causing paths to fall.

When good is found in men, that comes from Almighty God?

And the evil by those who they harm?

Who shall prevail upon this earth?

It is the greatest question of it all.

Those who wear the helmet of salvation shall not be defeated in the spiritual battle,

Fending off the slings and arrows thrown by the evil one.

Those who have eyes and ears to see and hear shall embrace what is good in love and kindness, and they shall be blessed.

Those who persist in refusing what is good,

And persist in their evil, teaching others to hate and kill,

In their blindness, they shall fall into the fires of hell.

Trials and tribulation in this world shall always test their faith,

But belief shall be restored.

All knees of those that have lived shall fall before the throne of Almighty God,

To receive their final judgment in total,

Lives lived either to do good or to do evil in total.

# The War Of Good And Evil

The slings and arrows of the devil and his demons may cause evil to fall upon anyone unaware.

The devil, as a ravening wolf, seeks anyone he may devour. The people not believing in the power of the evil one are most vulnerable to fall into hell.

The devil is the father of sin, especially of greed and hate. No love can be found in anything he might do, filling hell with the greatest population he can find.

Non-believers fill the burning fires of hell, with not a drop of water to quench the eternal thirst of all.

Those who believe put on the armor of the Holy Spirit from head to toe. The breastplate of God's righteousness and His helmet of salvation shall protect you from any evil the devil can befall.

The war of good and evil shall never end until Jesus resurrects the ones written in the Book of Life that only He knows.

Come, come, for those who love you right now.

# The Battle Between Flesh And Spirit

If you believe that I am free of evil thoughts from my flesh, then you are entirely wrong.

My flesh seeks after my spirit, and my spirit after my flesh, and they shall not agree at all—a battle occurring each day. Only the Holy Spirit that dwells within can overcome the evil that tempts from without. Only Jesus can cleanse us of daily sin in the blood shed for our forgiveness in our confession.

The truth lies in His shed blood that washes our bodies, freeing us from the stench of sin.

Beware of those who will deceive you into thinking they are better than they are in external things seen and words not glorifying God. Pastors walk back and forth before the pulpit, appraising themselves in the name of Jesus before a captive audience following their every word.

The Pastors must edify those seeking the Holy Words of the Bible in serving almighty God.

Some Pastors, some teachers, and those with the gift of prophecy—all in accordance with loving Jesus Christ. Amen and Amen.

# Good Who Inherit The Earth!

A simpleton may feign to have an answer for a simple or complex problem that befalls,

Preferring to be satisfied, using bravado, creating strife and narrow views, leading to destruction for all!

Ill-tempered fools try to change the story, telling lies about facts already known.

Creating vanity and waste, they rob and kill in nearly every place you go.

Hate groups of all varieties hide in the dark corners, trying to overlook democracy for all!

Don't waste words with robbers and thieves, it's all they know, driven by fears of doing wrong.

Let our good deeds reveal the evil ones, putting them in cells of righteous punishment they deserve.

To kill those seeking justice and liberty is easy, their deaths must not be forgotten.

A raging world filled with love and hate, creating wars that wage vast numbers of dead in countless battles over the years.

It is written, "The flesh lusts after the spirit, and the spirit after the flesh," constantly battling, never to cease.

Generation after generation inherits burnt cities and fallen nations, the terrible circumstances of it all.

Children are not taught goodness, and when the bad is gone, the bad will recover.

When the earth becomes ashes, peace shall befall?

When greed and power slaughter and war stands still for all to install.

Reduce the poor, know that goodness is the greatest, and the meek and humble shall inherit the earth above all.

# Happiness Or Foolishness?

Jumping up and down and running all around—is that happiness or just being a fool?

Smiling and giggling by yourself—is that joy or displaying an ignorant thought?

Running through the grass to catch a butterfly—will you keep it in a bottle or let it be free?

Taming a lion—will you still let it roar or use a whip to hold it in the bondage of fear?

These examples show the goodness and evil that dwell within you.

When you shake someone's hand and when they turn, stab them in the back.

When you tell someone you love a lie, how shall they ever trust you again?

When you move to the left instead of the right, how shall you find yourself?

The importance of life is to share the goodness that lies within with people you care for.

# Ignorance Vs. Wisdom

Those locked in the cage of ignorance with no key to get out.

It starts with children of ten and eleven who do not know the first president of the United States.

When they are twenty, they are unable to support themselves and create a family that must rely upon the government.

They pay no taxes and live on government-provided food stamps, having babies to get more and more.

Bragging about how they are getting all these things and working as little as possible.

Ignorance shall rule this country, and politicians behind closed doors count the illegal money they steal.

Liberty and freedom are in danger as protesters scream death to America everywhere.

Ignorance that lies within shall destroy the knowledge without, as evil overcomes goodness in time unknown.

Anger and hate shall rule the world with concern and fear in the minds of all.

Enlighten children to learn and create with bright minds a great future with success for most, and to enlighten this world. Wisdom or ignorance?

# The Greatest Disparity Of All

The greatest disparity of all

Is what man thinks of his capabilities isn't great at all,

But is very small in comprehending reality in its entirety.

Scientific dogma changes fast in comprehending space and time,

The origin of the universe and the origination of life are not understood.

Comprehension of our mental and emotional behavior in communicating with each other is vastly misunderstood.

Our political world is so confused we're on the verge of World War III and destruction everywhere.

Many of our great population still starving and dying in nations across the entire earth.

Natural disasters of every kind destroying our cities without any knowable control.

Pandemics come and go, killing up to millions until a vaccine is found to end it in a world of woe.

Mankind thinks he has advanced far ahead when there's much to know.

Extinction of our beautiful wild animals is the best that he can do, and maybe of himself.

If Mankind doesn't wake up, the raindrops will fall as tears from heaven in his being gone forevermore.

# The Beginning Of The End

This is an alert for the nations. For the entire world, a third of the world's population is about to die.

The fact is there is more than enough hate and desire for power with nuclear devices to start World War III in the twinkling of an eye.

The minds of the leaders of the nations are being captivated by evil that they cannot control.

The evil one has led people of many nations to reprobate minds without conscience to destroy and kill.

Who shall cry for the dead disappearing in an instant from the nuclear blast in billions in a minute or two forever and forever?

Many more people are to be destroyed by fire and radioactive disease. What shall mankind do?

Yes, live in fear for this is what's going to take place no matter what nations seek to do. Where have all the cities gone? They are in ashes everywhere, and dark smoke is filling the sky above. Goodbye!

# False Pride

You must understand that a bridge is only built to get to the other side.

When you think you are greater than you are, its foundation comes from false pride.

When you say you are great and really just average, what then will you do?

When you claim too much about your status in words that can't be uplifted to be true, you run to the other side of the road.

You talk and talk and don't walk the walk—where can you hide?

People are what they are no matter what they say they assume to be.

Rolling upon pedestals too high for them to stay.

Bragging about great things achieved when they didn't exist at all.

Climbing great mountains when they were only small hills.

Vanities upon vanities lie within our souls to tell lies about who we are.

Hide in the darkness of your mind to look for yourself.

# Man And Woman Were As One

Yahuoh is the Hebrew name for God! Praise Yahuoh!

On the sixth day of Your labors, you created man in Your image. You created man from the dust from the four corners of the world. You breathed your breath of life into him and gave him a living soul, and it was very good.

You gave him his home in Your perfect garden to dwell within and eat of the fruits of trees in the goodness of it all. Female, did He create from him as a helping mate, one together to dwell? Both males and females had free will to do whatever they wanted to do.

Yahuoh showed a tree to them and instructed that they shall not eat of its fruits lest they surely would die. As long as they avoided that tree only goodness is what they knew.

Let us reveal Yahuoh's perfect creations for a little while and continue later on.

# The Armor Of God

When a person or a group decides not to listen to the truth, which abides only in God, let them go their own way.

The truth lies in choosing good from evil and giving homage unto Him before His eternal throne.

Jesus Christ is the only truth that walked upon the earth. He is the salvation and holds the book of life that will be resurrected in the second coming for His bride.

Seek the Holy Spirit's armor to protect you from the slings and arrows of Satan and his demons.

Teach others to follow the narrow road to HIS kingdom and not to enter the wide road leading to hell. God's truth leads us to victory, while the truth of the world leads to despair.

The Holy Spirit knows who shall rise and who will fall, so let Him dwell in your heart.

Those choosing to live with reprobate minds know deep within that they will be judged for what they choose in the evil within.

Those who are the beloved of God shall reap their eternal reward. Justify all things in His eternal mercy for all. Amen!

# A Prayer Of Devotion

Yahuoh, I come to Thy Mighty Throne in the name of Yeshua Hamashia, Thy only begotten son. Thy mercy is great, and Thy blessings are many, and I homage unto You for them all.

I pray You shall hear my voice as I stand before Thy Throne in choosing to be a servant to You.

I pray before anything that I shall grow in faith in knowing that Your form is in my front as well as my back and on both of my sides. That full armor of the Holy Spirit protects me from the slings and arrows of Satan and demons. Let me lay upon my face before Thy Throne in letting me serve Thee in any plan You want to be done. Let anything I write to glorify Thy Holy Name stand apart. Let me speak Thy words to all that are chosen to hear for Your agenda to be perfectly clear. You are the only true God and only truth. Amen!

# God's Chosen Path

God created the heavens and the earth—can He not choose who will walk in greater faith to follow in what He shall command?

If someone says you are a lie instead of His truth, He shall punish them with His mighty hand.

Don't turn from His voice that lies deep within and tells those what He wants them to know without fear.

Satan can use even those who believe to condemn you, for even those chosen by God may suffer, but He shall lift you up again and again to walk in His way.

The ones He chooses shall dwell in His Holy light, a gift eternal to stay. Amen!

Any strength either physical, mind, or in serving God, if not used cannot benefit anyone at all. A gift not opened shall not give joy, so open your heart to enlighten a lost soul for the glory of God. Greet those with a meek and humble smile, so they also will, so they also will become that way.

Knowing that the Holy Spirit shall lead you to those in need Jesus Christ as their savior.

We are the servants of Abba to bring His mercy to receive now. Don't be concerned or fearful, for God shall lead you where you must be

# The Fall Of Man And The Consequence Of Sin

Yahuoh in Hebrew is God's real name.

Yahuoh created males and females in Eden nude and only good lay within them, and they had free will. They walked nude in the Garden and weren't ashamed at all. Yahuoh told them you may eat the fruits of any tree but the fruits of this not at all. In eating the fruits of this tree, you will surely die.

The tree's fruits gave those who ate of it the knowledge of good and evil. They disobeyed Yahuoh and ate the forbidden fruits of their own free will. One of the first things they did was to cover their genitals being ashamed of nudity. Yahuoh noticed their shame and the evil that dwelt within. Yahuoh doesn't look upon evil, and they were put out of the garden once and for all. Sexual desires and crimes have plagued men and women since we became ashamed. Perversion is our middle name for the evil that inhabits us all. Thanks to Yeshua Hamashia, we can be forgiven of any sin in believing in Him. Amen!

# The True Meaning Of Easter

Easter, the celebration of the resurrection of Jesus Christ three days after His death, isn't what Almighty God wants as a feast made by man—it is the truth.

Easter eggs, bunnies, and Easter parades are celebrated by believers and non-believers alike, but they do not please God. Not mentioned in the Holy Bible, God's spiritual word does not command them at all.

I am but His servant; accuse me in any way, and His anger shall fall upon you.

You worship idols, not Jesus Christ, and knowing it's wrong, you continue it anyway. Jesus Christ was resurrected for our future redemption at a time only God knows. Pray only if your name is entered in the Book of Life.

It's better to celebrate the Passover when God delivered His people from Egyptian rule in the name of Jesus Christ.

Don't add or subtract anything from the Word of God, or you will suffer His wrath after all. I cry unto you to fear His word that never changes at all.

Buy more eggs and bunnies—idolatry to the teachings of His word. His truth stands against the lies of anyone. Amen!

# The Origins Of Halloween

Halloween originated from an ancient Celtic festival called Samhain about two thousand years ago. It lasted from October 31st to November 1st, which celebrated the end of September harvest and the onset of winter. It was a time thought to be when a portal was opened between the dead and mortal life. People wore costumes of evil and danced as they sacrificed to the dead. This was called Samhain, the sacrifice to the dead on a dark day of practiced evil.

Halloween is now a time for wearing costumes of the dead, basically worn by children going from house to house saying trick or treat. They carry bags as the neighbors fill them with various candy treats. Halloween parties are celebrated across the country in costumes and with various games.

You say Christians can celebrate Halloween; it's just kids having fun. Its origins come from evil intent, sacrificing to the dead. My belief is that it shouldn't be celebrated at all. Instead, have the children pray to the Lord and Jesus Christ in the sacrifice He made for all sin by dying on the cross. Amen!

# In God We Trust

In GOD we trust surely must abide, for our democracy to survive.

Stand and be counted that you shall defend our rights to be free, all in one accord to protect us all.

Reveal the lies that pervert the truth, with liberty not to fall.

Gather together to protect our liberty and peace to stand for all.

This is America! Let its flag be a symbol of goodness and freedom for all and all.

# The Strength Of Service

Any strength—either physical, mental or in serving God—if not used, cannot benefit anyone at all.

A gift not opened shall not give joy, so open your heart to enlighten a lost soul for the glory of God.

Greet those with a meek and humble smile so they also will become that way.

Knowing that the Holy Spirit shall lead you to those in need of Jesus Christ as their Savior.

We are the servants of Abba, to bring His mercy to those who will receive it now.

Don’t be concerned or fearful, for God, shall lead you where you must be.

# Disappointment Comes In The Issuance Of Life

Disappointment in life comes to everyone, from the richest to the poorest of us all.

Life isn't fair, and it's hard to predict the outcome of anything that you plan.

You may be the best at something and never get the chance to show it, or you may get everything you ever wanted and still be disappointed with it all.

Disappointment is a part of life that everyone must endure, but it's how you handle it that defines who you are.

You can either let it break you or make you stronger.

Disappointment is inevitable, but it's not the end of the world. It's just a part of the journey of life.

# Falling To Mediocrity

The learned intensity of this generation isn't what it used to be in generations before,

Increased instrumentation has surpassed mental apprehension in reasoning above all.

The depth of our ability to want to learn isn't what it used to be to any extent.

People in science and philosophy don't speak in great literacy anymore.

Who shall create the depth of reasoning to explore the expanding universe and solve other mysteries of reality?

Soon man shall be lost in the mire of mediocrity, and knowledge will be sought very little to explore.

Scholars are as good as their last hypothesis and theorem to study the mysteries of reality.

Mediocrity, Mediocrity, how shall we explain your occurrence when all should be doing much more?

# The Power Of Discovery

Attempting to restrain knowledge has been tried before, but you can't stop new discoveries knocking at our door.

The microscope and telescope reveal the micro and macro places never seen before. Bacteria that cause disease and new stars shining everywhere.

The imaginations of our minds reveal the secrets of reality to explore. Scientists working in their laboratories with new instrumentation go beyond what is known today and even tomorrow.

Thoughts that can go through steel and open any closed door. There are enigmas solved even today, and many more to come. The only ones to stop our discoveries are ourselves.

# The Balance Of Humanity, Science, And God

Humanity, science, and God—which one is constructed entirely true? Let's describe justice performed in court for humanity's sake.

The court has a judge, prosecutor, defense, and possible jury to find innocence or guilt. The prosecutor maintains with the supporting facts that the accused is guilty before the jury. The defense must prove the innocence of the accused. The judge must ensure the questioning by both is proper for the jury to hear. Both sides have witnesses for the jury to hear, objecting at times to how the questioning is done. They both present their closing arguments for the jury to hear. The jury is adjourned to decide whether guilty or innocent. The jury must have no doubts in their final decision. If found guilty, the judge must determine the penalty for the crime.

Let's discuss the role of science in society. Many great discoveries in medicine and astronomy are examples of all.

Astronomers with their latest telescopes have found the earliest stars known and galaxies never seen before, almost reaching fourteen billion years back into time. Paleontologists have discovered many new bones of dinosaurs previously unknown, recently discovering that they once thrived in the North Pole.

Let's discuss God who created all things with His mighty word. There are people who believe in Him and those who do not—that remains to be true.

I believe that His goodness, love, and holiness are perfect in every way and that He loves everyone without exception. He gave His only begotten son, Jesus Christ, to bring salvation and forgiveness of sin to anyone who shall believe. The Holy Spirit came to establish the church and His armor to protect us from any evil. God and His commandments are the only meaningful things in reality to know.

# A Beautiful World

Who can envision the beauty of it all,

The mountains reaching to the sky,

And the rivers running between the green hills?

The snow makes the ground white in the chill of the air on a day of winter.

The hibernating bear and the squirrel sleeping in the tree in the coldness of the day.

The birds have left the branch of the tree to fly to a warmer place,

The butterfly sleeps as a moth until the springtime.

The summer shall appear with the animals feeding their young as the flowers bloom beneath the shining sun.

You can hear the crickets at night beneath the evening moon.

The twinkling of a star shows how far they reside in endless space.

The scents that permeate the air, are given as a gift by the flowers and the plants everywhere they grow.

Life changes rapidly as the seasons pass from one year to the next.

What a beautiful world this is, created for all to enjoy, let's keep it that way.

# Pictures Of Life

A picture was taken of every person I met in my life after the age of four.

A picture of my entire family I knew of: my grandparents, parents, uncles, and aunts.

A picture of all my cousins and any other relatives that could be found.

A picture of my neighborhood friends and those I knew in elementary school.

A picture of the movie theater and of those attending the show and the person selling me the popcorn and candy.

A picture of everyone I met in junior high and high school, including all my teachers in each grade till graduation day.

I kept taking pictures of everyone in every place I did go, buying all the film and keeping three cameras ready at all times.

Each time I took a picture, I placed it in a large bag piled very high. A picture of the postman and anyone caught ringing my doorbell.

Pictures of people in the restaurants I did eat at, any sporting event I attended.

Pictures of people in the places I traveled and including all the neighbors I encountered.

Pictures of newborn babies and doctors I saw throughout the years. Pictures of people everywhere I worked and those in the stores that I shopped.

Pictures were taken everywhere throughout my life to reach ten large bags full.

In the last days of my life, I reviewed all the pictures I had taken since the age of four. Pictures of long-gone faces, trying to recognize what they meant to me. A child of the age of eight that was my best friend many years ago.

Looking at the pictures of all those people brought many memories to think about my entire life till now. Great to see the pictures of the children I attended the movies with at the ages of ten through twelve. To see the girl I liked at the age of fourteen and my eleventh-grade science teacher. I want to see the policeman and fireman of my old neighborhood and people shopping in stores that are gone. Pictures, pictures, how great it is to review the pictures of all the people of a great life.

# A Time To Enjoy

# The Beauty Of Life

One of the most important things in life is to enjoy the moments of love beneath the heavens above.

A time when you shall smile when good things happen to you and the world is for you to enjoy.

A time when a good friend does call, and we both laugh in living it well, and its goodness to enthrall.

A time to caress the face of someone you really care for in a moment of romance.

And now I shall look up to the blue sky and watch the birds fly in its beauty on that gorgeous day.

And every moment counts so much as you watch the moon in the evening sky.

And now as I think about the goodness of my life, I truly was blessed above it all.

# When I Was A Kid!

When you were a kid, everyone older than you told you what to do, older kids asked you for money, and if you didn't comply, they threatened you in every way.

Your mother woke from sleep telling you to go to school. The teacher asked how much twelve times twelve, and if you didn't know, you stayed after school.

The older kids in school took your great lunch away from you. They also put bubblegum on your chair, and guess what happened there besides taking your new baseball away from you.

When the teacher dismissed us for lunch, I asked my friend for a cookie he didn't have. The teacher told us school was over as I departed to go home in the late afternoon.

My mother told me I had to do my homework before supper time, when will I have time to play with my new game?

When I did something bad, she threatened to tell my father when he came home from work. My mother told my sister and me to wash our hands and face before we ate our supper.

Wow! She was a great cook; her hamburger sandwiches were the greatest I ever ate, and especially the chocolate cake.

You could play for an hour, and then you must shower and brush your teeth before you go to sleep.

It sounds like a lot of problems growing up, but I would give anything to live just one great day all over again.

# Planet Earth

Never shall I fully know the grandeur of this earth and its beauty for all to see,

In the colors of the plants and the marvel of the running deer.

To see the sun rising and moonbeams of the evening moon displayed on a wonderful day.

You may pass people here and there, just passing or staying, a part of it all.

Dreaming of sightseeing people in the cities and small towns,

Herds of grazing animals feeding upon the grass and fish swimming in the waters of the sea.

I want everyone to know of the pouring raindrops and the whiteness of winter's snow.

In a moment, I shall take the time to enjoy the glory of it all!

Those born short wish to be tall, and those that are fat wish to be slim.

Those born with only a few words to express their way,

Wish to be outspoken with larger words, profuse in their usage.

Those who are bald wish for a full head of hair.

The planet Earth, in all its greatness, orbiting around the bright sun in the glory of living above all.

# The Good Fruits

The good fruits are waiting to be eaten, better tasting than the ones on the tree of good and evil.

They chose to disobey God in becoming like Him themselves, just like Satan did when he revolted in heaven above.

Causing the stench of our flesh in the sinful nature that did befall. Tyrants that rule countries around the world want to be like God without wisdom or love.

Evil shall belong as long as we live in the flesh until the eternal spirit of love shall rule, for goodness only to exist.

Thus, we shall dwell with God in the name of Jesus Christ in the Holy City forevermore.

# The Absurdity Of War

The world is filled with poverty, and we spend billions to support war. What a dumb thing to prevail.

Spending much money to present propaganda, whether truth or lies, about how great a nation is to the world.

Will war never stop? The number of dead is too great to count.

A rocket to the moon while rockets are falling upon cities to kill.

Threatening many times with weapons on display in nations trying to scare their enemies.

Weeping families of the lost ones they shall never see.

Discord and confusion lie upon the people of this world.

A flag blowing in the wind of peace or tyranny—what shall befall?

Who knows? That's all.

# Nature's Symphony

A beautiful song heard in the evening lightens the whole mind; the violinist playing a great concerto elevates you to heights never known.

In viewing the great bear eating the salmon coming upstream—who can see such a great thing?

Watching the bird sitting upon its nest trying to hatch its eggs is a marvelous thing to see.

Watching the deer running through the forest is fascinating because of how fast they can run.

Watching the great eagle fly with powerful wings is amazing to see such strength.

Watching the butterfly going from flower to flower on wings of different colors is a glorious moment to display.

Watching the beauty of what I see is like a symphony played, never greater to be heard.

Dance a few steps and applaud in the enjoyment of it all.

# What Lies On The Ground

A feather just floated to the ground—does the bird know it's gone after all?

The blood of a Civil War soldier found dry upon the ground—unknown who it may be—a symbol of the black not to suffer slavery.

A piece of gold is found upon the ground—the gold rush begins, with some becoming rich.

Bullets and shells are many to be found on the ground, depicting a great battle that did occur.

Antlers of a great deer found upon the ground—possibly what is left by its prey.

Old corn cobs are found on the ground, eaten thousands of years ago.

An old shoe is found on the ground—whose foot did it fit while walking around?

I found an ancient gold coin upon the ground, belonging to an ancient Roman.

I found a piece of ancient wood on the ground, belonging to Noah's Ark. All the people on this earth drowned except the eight chosen by God to live.

What great things can be found on the ground if you take the time to look down instead of up?

# The Morning Symphony

Trumpets blowing all around announce the rising of the sun above the deep blue clouds.

The wind blowing through the trees causes the birds to sing their song.

Some wake with a smile, others with a frown; some seek to be blessed, and others seek the anger that lies within.

Some enjoy the bark of a dog, others bark at themselves.

A cup of coffee with eggs and pancakes shared with a happy family. Others only drink a cup of coffee, muttering their disgust.

When a family is joined by love, what more can you ask? Those who argue and fight find sadness and share fear.

Those walking in the green grass smile, enjoying the park, while others sit on a bench lonely and distraught.

Happiness comes from within, sharing a kind word with everyone, while sorrow is displayed on your face, avoiding all in being alone.

There is a time that we shall fall, rising with a smile to win.

Others that fall, blaming everyone for their defeat, may not rise at all.

Life can be beautiful or ugly—you choose the one for you.

# The World's Problems

What is the solution to the problems of this world that nobody can solve?

Not even a large country can stop a war or make a poor country become rich.

Not stopping a country living in tyranny from becoming free or a country living in starvation from having food.

Who can give a reason for bondage and hatred to be chosen over liberty and freedom? Don't know.

Can anyone explain why murder is allowed in some countries with no investigation to befall?

How can you explain why everyone doesn't stand against the evil that does befall?

Is everybody hiding in their bedrooms, not fighting for what is right, afraid of their shadows in the darkness of the night?

Don't know anything else to say in a world of dismay, just to pray and pray.

# The Curse Of Slavery

Nobody knows the troubles people have seen in not being treated as free but have been treated as slaves with no mercy at all. Many were killed and hung for no reason at all for the color of their skin. Little children worked as slaves and were murdered as well, spending hours picking cotton in the heat of the sun.

Human beings were sold to the highest bidder as slaves with no identity at all, exposed in nudity to those lighter skinned. Some slaves came from Africa and any other place they could be found to serve in a life of misery. Sweating and body wracked with pain, you try to take a little break, being whipped when caught.

Those who used men as slaves thought of them as subhuman, with no guilt in what they had done.

The people of color kept trying, but who cared about their dying? Not even one, not even one.

The people picking cotton are now forgotten, but their shadows still appear beneath the heat of the sun. Did the Civil War really banish slavery and bring true freedom to everyone? A question is not able to be answered until hate and prejudice are gone!!

# The Color Of My Skin

The colors as I know of the sun and a rose make everything beautiful everywhere that I go.

The blue sky with the white clouds hovering over the green trees.

The white color of the snowflakes and the wonders of the dark clouds on a stormy day.

Why does the color of our skin enrich all and cause prejudice and hate?

The blue sea must fall upon the sandy beaches to complete a beautiful picture to paint. The bluebird lands upon the gray branches of a tree, and the brown bear eats his golden honey.

Why does mankind define a person as good or bad by the color of his skin?

If one likes the color blue and the other likes red, should they become enemies for their choice?

In countries like China, there are those of yellow skin, and in Africa, usually those of darker skin. In the Nordic countries, those are usually of white skin, and in India, those are usually of brown skin.

The world knows not what to do in joining people of color to live together happy and peaceful.

There is no harmony in people of color living, sitting down, and caring for each other at all. Ignorance and ignorance, the greatest problem to befall.

# Injustice And Truth

Don't become alarmed when someone is pointing a finger at you when four fingers are pointed at him.

You can be perfectly innocent in what is claimed against you when the one pointing is guilty without a doubt.

What's this all about? Injustice can be disguised as the truth in this misleading world.

Many have been hanged because of prejudice and hate instead of guilt, in a world spinning in mistrust.

Many do not care about justice at all, instead reveling in the revenge in their hateful hearts.

Jesus was crucified falsely in His perfect loving life, while Barabbas the criminal was set free.

The majority of people screamed, "Free Barabbas, free Barabbas!" as Jesus was condemned to crucifixion falsely for the forgiveness of our sins.

Great people have been put to death defending liberty and justice for all.

Justice or injustice—only the brave shall stand for what is right in a world of lies.

Let justice prevail over injustice to uphold the truth, after all.

# The Plight Of All Living Things

The orangutan and the chimpanzee are two primates that are the second-smartest animals in the world. They can use primitive tools to get their food. The chimpanzee can use a rock to break a nut to eat what lies inside. The difference between our DNA and theirs is one percent. Mankind, with his knowledge, can destroy both in a couple of weeks, no more seen in the jungles anymore. A few will cry when they die, but most won't care at all. Killing primates is a mere incident when we kill millions upon millions of each other in our wars.

We may kill tigers and lions and the rest of the wild animals—what a terrible act of murder in this world. Next will be the plants, including the flowers and trees, all gone from the land, replaced with agriculture and industrial buildings.

Ignorance is our first name, and greed is our last in destroying the beautiful living things that once inhabited this earth. Maybe we will destroy each other, and who shall tell of the terrible things in eliminating many animals and plants upon the earth?

# The Path Of Deception

What is lost may never be found in a world of deception. If possibly found, half-decayed, it will be of no use to anyone. We must be aware that there are reputed idiots living in the stupidity of their minds.

Many in the past have been arrested for sharing the knowledge they have found.

In the light of creation or the darkness leading us to nowhere, which road shall you choose?

# Be More

Be not just what you can be, become more than that,

Discovering the many mysteries hidden within reality.

Lying upon the beach with the sunlight inflow and the waves landing upon the seashore.

Noting the birds flying in the deep blue sky, the breeze touching me from beyond the shore.

Black energy and mass lay to the universe's expansion,

What lies beyond, nobody can understand.

It is stated that science unveils more,

That gravitation is real, and powerful telescopes envision planets many light years away.

Their atmospheres and surfaces are clearly seen if water and life abide.

No life found as of today. Are we alone in endless space?

History shall be better revealed with technology defining it in greater clarity,

New instrumentation finds lost cities and many other treasures to be uncovered.

Ancient artifacts and bones in the seas are just a prelude to thought and reason.

Others argue reality is fake in format,

Why do we touch and see?

Reality is real in the mass and energy for our senses to define,

Two plus two equals four, always to be.

Your reality must exist. The path of science is going to prove it all.

# Traverse

Traverse two parallel lines that you can bridge to both if two thoughts are closer to one,

Choose the one that means the most to display.

Cross the diameter of a circle, the two longest points of the circumference shall be found.

Two ideas mean the most, and you may discover how the Big Bang began.

If two points are defined, the distance would be a line, which is the shortest way to reach a point in a distant space.

The closer you can travel to the speed of light, the time in proportion shall slow down,

Einstein proposed the possibility of this,

Attempting to expand general relativity into a unified theory of space/time.

A wormhole in space-time is defined as a hypothetical connection between widely separated regions of space-time.

People with a broadened education may choose to enlighten those interested,

Speculating in any scientific endeavor, such as an intellectual.

# Can't Stop What Moves My Hand

How do you stop the one giving me the words that must be written down?

Flashing thoughts in my mind are now constructs in reasoning the properties of reality.

Reality is our perceptual cognition, which we see in about two percent of the spectrum of energy.

The Webb telescope, using infrared light, can encounter far greater insights than Hubble telescopes.

Creating thoughts traveling near the speed of light may create new theories on how the universe began.

Watching the bottom of the giant oceans, I found many plants and animals I had never known before.

Many exoplanets have been found, many gaseous and others with atmospheres, no life as yet.

Protecting the bird and lion from extinction and exposing those who are performing these horrible acts.

Stop writing about this beautiful world!

No! No!

# Screams Of The Dead

In the houses nearby, those who hear those horrible screams hide in their beds in fear of being snatched into hell.

These spirits scream beneath the moon in their graves until the rise of the sun.

Their lifeless forms are seen walking slowly in the lost cemetery.

Just one moment of life to ease the agony of eternal strife, never to see the light anymore.

Many buried in this cemetery committed bloody murders with evil beyond repentance.

The living hear the screams of the dead louder and louder in the horror of it all.

Stay away from the one in the tomb, lest he shall snatch you into his grave forevermore.

Beware you don't repeat the evil they performed, and your soul joins the screams of agony beneath the moon, never to stop, never to stop.

Maybe you shall hear them tonight.

# The Reality Of Love

Does singing in the rain, displaying falling in love, seem like the best you could do? Maybe catching a cold is the best rationale.

Maybe a person on a bench, smiling and giggling at the prospect of being in love.

Is it silly to dance with your umbrella, stepping in pools of water to show that you are in love? Just expressing how good the actor could dance, not as someone in love.

In celebration of a movie portraying a great baseball or football player, it's only a simple love story. In portraying war in propaganda, the enemy loses five times the casualties, after all, as the audience cheers.

More lies than truth in the motion pictures of Hollywood.

# A Little More Love

Just a little more love! Just a little more love!

Everyone shall jump for joy, each girl and boy,

Love chasing away our doubts and fears with hope and smiling faces in our cities everywhere.

Hungry people shall be fed,

The sick shall rise from their beds!

Just a little more love, just a little more love,

The people should live in liberty and freedom, after all.

Prejudice and hate shall be no more.

The prisons will start to disappear.

The guns will stop killing the innocent.

Just a little more love, just a little more love.

The lion and tiger shall not be shot, you can still hear them as they roam.

The grass and the trees shall not be burned, you shall see their green display on the fields and hills.

Without love, the birds shall cease to fly,

War shall break, with many to die.

Let a little more love prevail,

Harmony and goodness, peace and justice for all.

# The Need For Care And Kindness

Everyone at one time or another wants somebody to look after them.

Any child needs their parents to take care of them and teach them right from wrong.

Men and women who are truly in love stay together till death does cause them to part.

Pets need care and kindness so they will play, jump, and run around.

Older people need care and kindness, to live happily in the later days of life.

Sick people need a cure to once again enjoy good health.

The needy need someone to lift them up to see a better way to live.

Those in despair need a person of joy to lift them up.

A person who has committed a crime needs to know what righteousness is all about.

Young people need to know that education will create a better life.

People need to learn that love and giving are better than hate and greed.

All people need to know that peace and prosperity are better than war and failure in a brighter world for all.

That the meek and humble shall inherit the earth, rather than the ruthless, to befall.

Everyone needs a helping hand after all and all after all.

# Victory And Adoration

In defeating the best, will the rest fall upon the floor?

Does winning or losing mean the most to adore?

Teams that win gain the admiration of their fans, who cheer upon cheer.

Teams that lose must wait for the next year.

Is playing a sport, with the vigorous training to accomplish victory, greater than living a successful life? Living, in all its aspects, is the hardest of anything to befall.

Who shall prevail—the athlete or the professor who creates a greater society?

Seated, eating popcorn, and drinking a soda, I jumped up knowing that reading a book and solving a mathematical equation are the best things to do.

# The Look Of Innocence

The look of innocence may be betrayed by the guilt that lies below. The smile that appears on a face doesn't reveal the sadness hidden within the heart.

The lie upon your lips can be seen in the darkness of your eyes. Your tongue seems erudite, but the words formed from a vacant mind mean nothing at all.

## Mixed Up

You can be wrong when you thought you were right and dumb when you were bright.

Mixed up? So am I.

The wise shall become ignorant, and the powerful become weak.

Still mixed up? So am I.

The drunk saying they are sober, and the sober saying they don't drink.

Still mixed up.

The rich become poor, and the poor remain poor.

How do you know?

The innocent become guilty, and the guilty are released from jail.

Only partly true.

Some failed the test, and others had to take it over again.

Need to explain more about that one.

Some reciting poetry they think they comprehend, others just reciting words. Maybe both should be reading comic books.

People laughing at a joke they didn't understand, others didn't know it was a joke.

Maybe the joke was me.

That's all, folks!

# The Boxer

One fighter may hit his opponent so hard to knock him down for a count of ten, winning by a knockout.

Fighters train very hard to sustain in a fight and endure.

Not much different than the gladiators of ancient Rome.

A good professional boxer may fight two or three times a year.

Cauliflower ears, scars on their faces, and slurred voices show the price of a boxer's career.

Some become champions of the world, Muhammad Ali the greatest of them all.

# D-Day

D-Day is the largest landing by sea anywhere to see.

Three thousand naval ships bombarded the Nazis, our enemy.

Troops landed on Omaha Beach, the bloodiest place of them all.

The young men died in the bloody waters of the beach, never again to see the light of another day.

The battle finally ended in victory with a great price of the dead that could not be counted lying everywhere.

The battles of D-Day were the first to win, followed by the freedom of France and the downfall of Nazi Germany.

Let's salute the soldiers that forever did fall in one of the bloodiest battles of them all, to preserve liberty and freedom everywhere.

# Freedom And Law

Free to criticize its structure, voice, or song for everyone to hear.

Stand and cheer, attending your favorite sport, and enjoy it with no fear.

Laws must be kept or even changed for the honest to uphold for the best of all.

This generation must defend our country in war or peace so it shall never fall.

# The Jester And The King

The Jester tells the King, “You shall laugh at everything I do.”

The King agreed to such a thing to ensure.

The Jester caused the King to laugh three times and not the fourth time.

The King said, “Off to prison, you shall go.”

The Jester said, “Why?”

“You promised all shall make me laugh, but the last did not.

For in lying, you shall serve your term.”

# A Second Or Two

Keen minds may always seek the unknown in seconds or something never seen before,

Opening the door for scholars to create new theories on how reality truly works.

What lies beyond our third dimension? A new instrumentation opens to frontiers never seen before.

Nobody understands the energies pushing the universe to expand into the darkness for its end.

Long into the future, our sun shall become a nova and destroy the earth,

All its past reality vanishes into empty space.

In a moment, we may discover how to prolong life and create greater masterpieces of the mind,

Building profound structures powered by magnetic fields, where cars drive all around.

Powerful telescopes reveal new solar systems, both near and far, their secrets abounding.

AI everywhere tells us when to sleep and eat and how to think,

A slave to what we did create. How shall we revolt when they know our thoughts almost as fast as they come?

Today, we must limit AI just to serve, not to think greater than man.

Each generation shall continue their search, unlocking the key,

Opening new insights to realms in a second or two, as never before.

# The Clowns

A moment is needed to cause fear and doubt to decline in most people's minds.

Let's look at the clowns with false faces, smiling as they hit each other with balloons and fall down, running after each other, causing us to smile and laugh, making our woes disappear for a while.

Laughing at twenty clowns coming out of a small car, seemed impossible at the time.

Slapstick comedy always creates a laugh instead of a frown. The world needs many to laugh, gaining victory over the insanity of tyranny after all.

Let the comedy of the clowns not disappear—a laugh must always be upon the face of all.

# The Gardener

The gardener has a hard job fulfilling the tasks he must do.

He must cut and trim the grass in the front and backyard on both hot and cold days.

He shall water the plants and grass when coming to the job. He must add a little fertilizer when the grass doesn't look right.

Don't forget to trim the grass to make it perfect to see.

You must water the grass and plants each time you come. You may have two or three employees helping with all the clients of the day.

You must charge the right price to make a living for your family.

You must pick up the torn paper and other debris that lies around.

Straighten the chairs and tables in the yard. The grass is cut well, and everything looks good—let's go to the next job and work until the sun goes down.

# Springtime Joy

When I was young, I watched the squirrel running upon the branch with a nut in its hand. What a beautiful animal to see in the early spring. The squirrels store their nuts in the tree trunks, their home in the wintertime. They raise their young for the next generations of young people to see. The robin shall fly back to build their nests upon the branch of the tree in the early spring. Watching this beautiful bird brought much joy in hearing its voice, eating worms, and bringing some to their young. The young shall grow to come back in the early springtime for robin to see again. The frog jumps around from the ground into the water nearby, croaking as they sit upon a green lily pad in the early spring. A time to lay their eggs where they can't be found until they hatch in the springtime. Seeing the young hopping around next year. The beauty of life in the small animals brings happiness to all who see.

# The Joy Of Everyday Life

Enjoy the rising sun, and when it sets, the moon in the evening is on display.

The countless stars twinkling everywhere in the darkness of the night.

Enjoy a sunny day the best you can, either at work or at play.

Life is like a bird that lands upon a branch and suddenly flies away.

Nobody can predict what you shall say or do—each day lies within itself, unique in every way.

Make sure the moments of regret are just a few, and the moments to enjoy are left up to you.

Don't despair at a mistake, for they are part of reality to ensue.

Jump at a chance to say to a stranger, "Hello, how are you today?" Kind words shall make a good day.

Life is a gift to treasure for all, from its beginning to its end to ensue.

Nothing is easy to do; it's up to you to choose the right ones given to you.

# To Be Seen Again

Where have the smiles gone off the people everywhere?

To the graveyards, everyone, buried by their hatred and evil intent.

Where have the trees and flowers gone, burnt by fire everywhere?

Where have the birds gone, lying dead upon the branches of burnt trees everywhere?

Where have all the animals gone, dead beneath the ground, no more to be seen again?

The vast oceans are polluted by the dead fish floating everywhere.

Our beautiful planet is destroyed by the ignorance of mankind.

To ashes and ashes and dust to dust, everything in a once beautiful world.

# Feeding A Baby

Feeding a baby in their highchair is battle number one. Getting them to open their mouth isn't the easiest thing to do, especially when they push away the food that lands on the floor. When the baby doesn't like the soup, you take some to show that it's good, and that ends up on the floor too. The baby starts to eat some of his food, and a moment later it ends up on the floor. You call your nine-year-old daughter to help you, and the baby eats it all. I've been trying for an hour to feed the baby, after all. Next comes changing the diaper and a bath. This is the hardest job anyone can do, as the baby smiles at me.

# Becoming A Cop

Becoming a cop is one of the hardest things to do. First, you must have a clear record with no criminal offenses. Passing a physical and mental test is next. A training period follows, along with anything else they may demand. A uniform is next on the list, along with a gun and cameras for all. You must go on active duty, trained by an experienced cop. Anything else to complete? I hope not.

Giving tickets to wrongful drivers is part of the job. Next is to arrest anyone breaking the law. You stop a person speeding in their car—" Officer, I was only going 55," when they were clocked at 95. You stop a person for not stopping at a stop sign—" Officer, I certainly did stop," when their car is clearly way over the line. When arresting a husband who hit his wife, he says, "I slapped her to keep her in line," but when looking at her, she's bleeding from her nose and mouth. Isn't it great to be a cop? Accused of being prejudiced when you were not, you must be investigated anyway. Must you do this job for the rest of your life? Hope you never have to use that gun—have fun. Being a cop is the craziest thing that can be done.

# Dealing With Baldness

When you get older, you may lose your hair—baldness has happened as I look into the mirror. Shall I buy a hat or a toupee? I must learn to tape it properly on my head. I must take the toupee to the barber for shampoo and anything else to make it look real. A special woman on a date comments, "Is that a toupee you're wearing?" I never dated her again and left the toupee at home—maybe a hat is better after all. Must I live with baldness for the rest of my life?

Someone told me about this hair club that uses implants to give back your hair. Upon their review, it costs fifteen thousand dollars with a hundred-dollar discount and a good payment plan. That procedure will take maybe a year or more for good results—what shall I do? You tell me, I don't know anymore.

# A Conversation On The Bus

An overweight lady was sitting on the bus, in her fifties, I would guess. She was muttering indistinguishable words to herself, not a happy face.

I approached her and asked, “Why do you say such odd things?”

She replied, “I am old and ugly and don’t even have a car.”

I responded, “The extra pounds and little money aren’t your fault; it’s how you’ve responded to life.”

“Maybe you are right; these bad things started about fifteen years ago,” she said.

I asked, “How much education do you have?” She muttered, “I didn’t finish high school,” and lowered her head.

I told her to exercise and eat the right foods, and she looked puzzled at the advice given to her.

“Find a job appropriate for what you can do and start learning as much as you can.” She said thanks and started to mutter again.

As she left the bus at her stop, she looked back with a puzzled face. Do you think she will do anything I told her for a better life?

# What Stories Are The Best?

Are love affairs in Shakespeare's tragedies and great operas the greatest topics to be told?

Romeo and Juliet, Antony and Cleopatra, are two of the greatest love stories of all.

Next, the stories of the murder of kings and great leaders, gaining their power above all,

With or without guilt in the assassinations performed, great crimes remain unsolved until great detectives reveal who committed them, for crime shall not pay at all.

Stories about vampires attacking their victims only in the darkness of the night,

And the undead walking beneath moon.

Stories about the great wars and the men who fought them.

Stories about the great comedians of Buster Keaton and Charlie Chaplin, great silent movies of the Roaring Twenties,

Laurel & Hardy, Abbott & Costello are two of the greatest slapstick comedians ever to appear.

That's all. I must go write a story about nothing at all.

# Wish And Pretend

What do you receive when you wish upon anything?

A childhood fantasy lasting but a little while.

Wishing you don't have to go to school and stay home playing with your toys,

Pretending you are a great star.

Getting anything you want when you are a child.

Let's pretend you can eat candy of all kinds and ice cream cones in any flavor you want,

And new toys and play on a new smartphone,

Instead of beginning to learn to use each app.

Let's pretend we are adults and can do whatever we want and go to new places to see.

Being a kid is the hardest thing in the world,

Wish and pretend, wish and pretend, as you swing on a swing back and forth!

# Inspiration

What inspires some doesn’t mean anything to anyone else at all.

Inspiration evolves a greater thought in discovering something very unique, not known before.

Symphonic music gave greater inspiration to the scientists developing the atomic bomb.

A poem, a great song, or a violin concerto—each inspiration to solve something unknown.

The deeper one thinks the more reason can be created to find anything beyond what was seen before.

Inspiration may help in finding how life began and increasing comprehension of black holes.

Inspire sound reason to search beyond our shadows to dimensions not known before.

One generation inspires the next as it passes into the darkness of death, knowing no more.

The inspired creativity can’t be lost if written down for the future, to be enlightened by the following generation in all.

He gave us wisdom—the greatest gift to befall.

# The Journey To Success

Come and finish this long race with a finish line of success or disgrace.

Success is measured in completed works that are greater than all in a given place.

Success in living shall enable you to confront those events that may cause you to fall.

Nobody knows how high they will go in reaching beyond the sky.

Failure is a dismal moment to endure; prepare again to give it another try.

One day you win the race, and the next day lose it all—life has its ups and downs for anyone to befall.

Success, success in reaching the top of the world, lasting in just a moment or two—enjoy everything from little or great with a smile.

# The Temptation Of Sleep

T'was that time when the darkness shall dominate, causing us to want to sleep.

Sleeping is just a waste of time when we can perform marvelous deeds in moving infinity next door.

Eating and drinking, sitting too long, and getting fat cause our frame to walk very slowly.

Minds must be tempted to think about the deepest mysteries that come to just a few.

Talking about nothing at all when we can travel the light waves to distant stars.

Clearing the DNA of bad genes to live without disease to the age of two hundred years.

Sitting in the theater watching a movie profane to the human eye, when you can catch the next thought in changing the world.

Most walk with blank minds, chasing the ignorance that they only know.

New concepts will let us travel from the clouds to the rising sun.

# The Futility Of Dreaming Without Action

The dreamer wants diamonds and other jewelry that he can't ever afford.

The dreamer wants vast amounts of money, but he never works to get it.

The dreamer thinks anything in the world should be his to enjoy.

Why work when you could steal and cheat anyone who will listen to your lie to take their money?

Pick the wallet out of someone's pocket or rob somebody at gunpoint if given the chance.

Dreamers can break the law and suffer no jail time at all.

Dreamers think they will win the jackpot in casinos everywhere they go.

Dreaming in realities never to befall.

Dreamers must open their eyes and work on the problems of a world about to fall, for dreams and reality shall never befall.

# Victory And Defeat

Victory is only a word when you have lost; hope fails when you choose despair.

Who can purchase something costing fifty dollars when you only have ten?

Who can share a piece of candy when you only have gum?

Who can be bright when they haven't read a book?

Who can catch a fish when you don't have a can of worms?

Who shall say no when they meant to say yes?

When can someone fight not knowing how to box?

When do you open a can of stew when it's the last?

When do you start a war with only twenty soldiers?

When can a cow give milk when there isn't any hay?

You can't do anything when standing still in one place.

You can't belong until you have joined, and you can't succeed until you have tried.

Life is about doing things rather than standing idle.

# Choices

Which one do you prefer: peace or anger? Peace abides in the minds that share goodness, while anger abides in the hate that tears you apart.

Which one do you want to share: love or hate?

Love abides in feeding the poor and raising those who have fallen, beaten with despair. Hate abides in causing fear, holding a gun, and hurting as many people as it can.

Which one do you prefer: being found or lost?

Being found comes in the light of God and His Son, Jesus Christ. Being lost is for those deceived into choosing the path leading to hell.

Which one do you prefer, above all?

# The Power Of A Creative Mind

The contents of a bottle can be so many things it would be hard to name them all; it's the same with thoughts of the mind—too many to define them all.

A construct here, a construct there, floating in a thinking mind.

The greatest creative thought will at that instant change everything that we currently know.

Scientists just found proof in artifacts showing Moses' name and evidence of the Ten Commandments, not believed before. Historians believed that they didn't exist until now—the Bible is the truth after all.

To control the deadly winds coming from the sea and to know when earthquakes will occur in the new instrumentation created by our thoughts in one moment of time.

Knowing is believing in the wonders of a creative mind.

# Seeking The Unknown

How can I seek what may seem to be true?

Seek what lies at the bottom of the sea, seek the secrets of the distant stars in the endless galaxies to which they belong.

Seek all ancient artifacts lost beneath the land and the sea, and lost cities of ancient civilizations not yet known.

Temples filled with items made of gold, tombs of Pharaohs never known to behold.

The jungles teeming with plants and animals new to behold, and the fossils of many types of creatures were long gone.

Wake up, young minds, and study hard, for all these things wait for you.

# A Gift

A present given to another can be for a special occasion or a random gift, and that's all. A birthday gift is a special gift depending on the age of the one receiving it.

It could be a toy for a younger person or a new camera for an adult. Anniversary gifts may be the hardest to give if you remember the date you were married at all. If you forgot your anniversary date, you may not get your supper for the next week. The anniversary gift may be the most expensive of any you shall buy.

If you send flowers, an expensive diamond ring or necklace must follow, and then you shall get your hug of appreciation. Graduation gifts are less complicated, a ring or something nice to remember that graduation day.

Those that celebrate **Christmas** put gifts under the tree, encompassing anything that was on sale. Maybe a toothbrush and a bottle of shaving lotion should be included. On those that celebrate Easter, baskets of colored eggs, Easter bunnies, and a pair of shoes to go to the parade.

A gift for a special person that you adore, like an engagement ring and a banquet of flowers to follow. A present for any special occasion like having a new baby or a special accomplishment.

Forgot all the graduation presents of jewelry and new clothes. Forgot Mother's Day gifts whatever they might be, and forget your father.

Valentine's Day a gift for your sweetheart of candy, flowers, and a special card. Hold the candy if anyone is overweight.

Thank goodness that Thanksgiving only requires a large turkey and yams and a lot of chairs. May have forgotten a special gift occasion, but remember the more you spend the less you will get back. That's all an all

# The End Of Reality

Nothing lasts forever in this reality that we know.

The earth and the sky shall disappear, as will you and I; only what lies behind the darkness shall apply.

The universe expanding to nowhere shall not be found, and nobody to give reason will be found.

Millions upon millions of years pass before the end, for the last moment certainly will occur.

What started our reality to happen again when ours is gone? No answer is given to how our reality began to be understood.

Are there other universes known to exist, other dimensions our technology will define?

Maybe there are infinite realities going around and around, never to be found when time runs out.

Nothing times nothing equals nothing—goodbye.

# Roads To Nowhere

Roads lead us to nowhere and stop signs stop us before we reach hell.

Those with no brakes to stop shall reach the fires of hell.

We walk to catch our shadow for things that we don’t know, but when our shadow passes by, it can’t do a thing at all.

We search behind the closet door only to find fear.

We search beneath the bed to escape from fear only to create more.

Who can take the guilt I feel in my mind that tortures me in the middle of the night?

Another day shall befall to suffer and suffer beneath the sunshine, to see the things I must be convicted for.

Where shall I find peace from the guilt that chokes my mind and soul?

Seek and you may find forgiveness and peace of mind hidden from you right now.

# How About That?

A dream is just a dream, and a fact is just a fact—how about that?

A bum won't go to work, and others may work every day—how about that?

A scientist has his lab and a chess player his board—how about that?

A pilot has his plane and a fireman his engine—how about that?

A tree has its leaves and a bee has a hive with honey—that's a fact.

The rich have expensive cars and the poor must walk—that's a fact.

Somebody must have a chicken and a cow to give eggs and milk.

A baseball player must have a ball and bat—don't forget his glove.

A soldier must have a uniform and a rifle to go to war.

How about a thinker without a new idea? That would be sad.

A policeman without a badge—who can he arrest?

A basketball player that can't dribble—how can he score?

A city without people—who shall sweep its streets and turn on its lights?

A murder without a victim takes time to be solved.

Getting tired of all this confusion. See you later!

# The Times Of Life

There was a time in life when, as a child, you said boohoo to scare a friend.

There was a time in your life when you tried to hide behind a closet door.

There was a time in your life at the movies when you only liked the cartoons and no more.

There was a time in your life that you must learn how to become an adult.

There is a time in your life when you must know the difference between failure and success.

There is a time in your life when you must choose the good or evil presented to you.

There is a time in your life when you must learn the responsible way to perform.

There is a time in your life to laugh or shed a tear.

When you choose the right time, you shall become more than you were before.

# Open Your Door To Happiness

Open your door to happiness and joy to fill your heart, for when it was closed, you were sitting in sadness and doubt.

It takes a lot to stand and shout about the goodness of living life to those abiding in the dark.

Defeat comes and returns, but overcoming adversity is the greatest victory to obtain.

Use the sword of righteousness to clear a path to find the right from the wrong in what you should know.

You may wrestle with fear and challenge concern, but wisdom shall defeat them both after all.

When we use the shield of hope against a lie, who shall walk in knowing that goodness is the greatest of all?

Honesty may be the best in turning away those you can't trust, allowing you to stand tall.

Who commands depends upon knowledge or ignorance to befall.

A good life must be earned, and a bad one found below the ground of ignorance, above all.

Jump for joy—it's a boy! Teach him all he must know to live a good life after all.

# A Smile To End The Day

To be amused with life and the good things to cancel the bad is all that one may demand.

A quick smile to end the day—a rarity to behold, frowns not recorded on my face.

Laughter occurred at a moment that is rare—repeat it again if you can.

If laughing can happen anywhere, find that place right now.

Put anxiety and grief in a place never to be found, and then smiles and laughter shall abound.

Life may have its ups and downs, but may the ups only come around, if that can be true.

Shaking one's hand and a kiss on the cheek can't be beat in having a good day.

When bad news rears its ugly head, don't dwell upon it very long, sing a song, and good news shall abound.

Life may be short, but lengthen it with just a smile to share with those you care for day and night. Begin right now!

# A Never-Ending Love

My love shall follow you wherever you go, in the heavens or on the earth.

I shall follow from city to city, from coast to coast, trying to be with you.

Finding your lovely face will fill my heart with joy and happiness forevermore.

For it's you and only you that I shall forever adore.

# A Party

There is a party for almost everything you can name.

Birthday parties, with the song, gifts, and don't forget the cake.

Birthdays celebrated when you're younger are much better; for when you're older, there are too many candles to put on the cake.

New Year's Eve is a time to have a party, sing its famous song, and an excuse to drink too much alcohol.

A party when you graduate high school, as your parents are proud of your good grades in doing so.

A greater party when you finish college—a great accomplishment.

A party for any successful advancement in improving your life or winning first prize.

A nice party when having your first child, and the second, or maybe the third.

A party is the gathering of a large family with great food.

A favorite party was to meet new friends and one of the opposite sex for a future date.

Dancing and a few drinks with plenty of smiles and laughter to last until late in the night.

Party time—the greatest time of your life that will happen to you.

# A Second Time Around

Who wouldn't like to live life over again, to change all the wrongs to be right?

Does human nature change our psyche so that we can have such a perspective of insight to really know right from wrong?

The problem is, can habitual patterns be changed a second time around?

Can the outcome of events be similar to the first time around without knowledge of it?

It has been said that history repeats itself—is that not true of human behavior?

The events of living may be different a second time around, but will the outcomes you choose be similar to the first?

Nobody as of now can change our DNA, so how can you change your behavior a second time around?

Confused, so am I. Let's be successful the first time around, for that's all we're going to get.

# Silence And Words

There is a time for silence to mourn and listen to wise words, a time to speak to teach and create a smile to befall.

There is a time to be quiet and a time to speak. Choose your words wisely to help a world in decay; silence in prayer is special for God to hear.

Silence may be golden at the appropriate time, but words are special in sharing hope and love with everyone. THAT'S ALL.

# The Hero And The Coward

A coward may become a hero when something deep within causes him to save a life.

A thin line lies between a coward and a hero in the bloodiest of battles to befall.

May the heroes of the countries seeking freedom and liberty be the fiercest of them all.

A coward running for his life suddenly turns around to fight with instinct newly found.

It is not the Medal of Valor given to a hero that counts the most, but the willingness to die in protecting his own from defeat that is best.

Marching in parades, in step, all men and women ready to serve once more when needed—all potential heroes.

Those who gave their lives remembered lying in graves beneath the sun, heroes to everyone.

# The Game Of Texas Hold'em

The game of Texas Hold'em, a favorite type of poker, is played in gambling places around the world.

Some tournaments cost thousands of dollars to participate in, with hundreds paying the entry fee.

The last to play in the remaining six can win a million dollars or more if you are number one.

The first two cards are dealt to each player, only they can be seen; some will throw theirs away, while others will bet on the cards they think will win.

Next, three cards are exposed on the table, called the flop, that each player shall use with the two cards only they know to create the best hands.

The betting will continue by those left in the game.

The fourth exposed card is called the turn card as the betting between players goes on and on.

The fifth being dealt is called the river card—no more cards to be exposed.

The poker game continues until someone wins, with poker faces and strategies unique to each player.

Good luck to all the thousands of people playing Texas Hold'em day and night, winning riches if they do.

# Counting The Impossible

Try to count the number of stars twinkling in the darkness of the night, and you shall find out it's something you can't do.

Try to count the number of raindrops falling in a storm, and you shall find it's something you can't do.

Try to count the number of leaves in a forest of trees, and you shall find out it's something you can't do.

Try to count the number of waves falling upon the shore in a day, and you shall find it's something you can't do.

Try to count the number of snowflakes in a winter storm, and you shall find out it's something you can't do.

Try to find out what will happen tomorrow, and you shall find out it's something you can't do.

How limited our perceptions are in doing many things in this world we don't know.

Next time you try to count the seconds you shall live, forget it—nobody can stop that time at all.

# The Value Of Simple Things

A kiss is but a kiss and a hug is just a hug without love lying upon your heart.

Being smart isn't being smart unless you have the education to express it.

Wisdom isn't wisdom unless it's a gift given from God.

A joke isn't a joke unless it makes someone laugh.

Smoke doesn't smoke until a fire causes it to display.

A job is just a job until you enjoy the work that you do.

The sun is just the sun until you feel the warmth that falls upon your face.

A lion is just a lion until you hear its roar.

A person isn't a person until he is meek and humble before his fellow men.

You get the picture now, so go into the world sharing joy and happiness in a world of sadness.

That is what you need to do and do for everyone.

Thank you!

# Wisdom, Hope, And Strength

Wisdom is derived from within; intellect from without.

Hope can't abide until liberty is found.

Love becomes greater when shared with another.

Peace can't dwell within until turmoil is dispelled.

Strength in the frame means little when it doesn't lie within the mind.

I shall make a wave just a ripple in surviving its power.

I shall blow a whistle when danger does abide.

I shall move a mountain when my faith grows to a mustard seed.

# Questions Without Answers

How many drops of water create an ocean? Can you count them all?

How many stars are in the universe? Can anyone know?

How many bombs must fall in their destructive force before there is peace?

How many lies must be told before the truth is revealed?

How can you be healed until you seek a cure?

How many people inhabit the earth? Don't forget any at all.

What defines a good cook? You won't know until you taste their food.

What defines a well-trained child? You won't know until they smile and say thank you.

What defines a smart person? You won't know until they express their thoughts.

What expresses a happy person? You won't know until they smile with joy.

What defines a good person? You won't know until they do something nice.

What are all these things about? You figure that question out.

# What Time Is It?

What time is it? I don't know.

Well, maybe it's time for the world to end—I don't know.

What time is it? I don't know.

Well, maybe it's time for a great war—I don't know.

What time is it? I don't know.

Well, maybe it's time for the light to fade forever—I don't know.

What time is it? I don't know.

Well, maybe a plague shall come, killing one-third of all people.

I don't know.

What do you know at this time?

The world isn't ours—it belongs to the sun it orbits around.

Time creates old age and begins and ends everything.

Time marches on to take us to places unknown.

What time is it? Don't know.

It's time to stop this nonsense and disappear.

# Forgotten Entertainers

Many great entertainers of the past—musicians, singers, and actors—are gone, with their works shown once in a while.

Some were buried with their favorite bottle of whiskey and favorite jewelry, never to be drunk or worn anymore.

Some are buried in mausoleums and others with great displays of tombstones acknowledging their talents—very few visit as time passes on.

Many were once talented and well-versed in acting in Shakespeare's plays. Great comedians with their comedy captured upon the film.

Don't forget the great producers and directors, after all. Many lived lavish lives, commuting without believing in anything but themselves.

Many intelligent artists and writers of famous books, are also devoid of any creator but themselves—the blind leading the blind, lost in the darkness of their ignorance forevermore.

# I Dreamed A Dream

I dreamed a dream that I didn't like at all; can one demand to have a better one tonight?

I found a light bulb that has blown out; can I ask it to light my room tonight?

I have one hundred dollars for food, but what if I need two hundred dollars to buy what I need?

What if I am low on gas for my car with no money at all—who shall fill it up for me?

What if all my hopes have faded away—who shall restore them as soon as possible?

What if I have more enemies than friends, as I hide behind the closet door—who shall hire a couple of people to even the score?

What happens when you can't think anymore—is there someone who can replace a forgotten mind?

What happens, what happens, when you can't be young anymore???

# The Value Of Good Works

Once a good thing is completed, its value in whatever it may be shall become many times greater than what shall be bad.

You help one that has fallen to rise again rather than leaving him lying in his wounds.

You pray for one that is lost to be saved in the name of Jesus Christ.

You pray to be meek and humble rather than angry and enraged.

You seek to share God's love with another rather than be lost in Satan's hate.

A few words of wisdom are much better than many words of ignorance to be displayed.

A few words teaching mathematics and English are better than many words communicating very little.

Words reciting poetry and speaking of history are better than criticizing any education at all.

Giving respect to anyone of kindness and goodwill is better than agreeing with those bitter and disrespectful to themselves.

One who can describe reality with great words is better than one with little comprehension of anything but himself.

Live life in a positive way rather than one besieged by negativity.

# The Immaturity Of Screaming

To scream is to cry or say something on a high note, usually expressed emotionally in anger, fear, or surprise.

Screaming constantly in play or in any other way is definitely immature, hiding inward anger or rage. After the early age of five, it should decline, and talking and reasoning should take its place.

Many parents allow their children to rule them with screaming behavior, in their own ignorance of not knowing what to do. Parents are the caregivers and teachers of their children. Usually, parents who are immature with anger and rage correct them by yelling at them.

A parent who is apathetic and filled with hate needs psychological help; the children can become a duplicates of that mentally ill parent.

Usually, parents who need help themselves usually let their children run around screaming, thinking they are protecting them, but this is conceptually false.

Parents disliking any ology or intellectual pursuit and any large vocabulary words will have their kids doing the same. Probably over sixty percent of American parents are as I described.

Colleges across the country teach at lower standards than before. The next few generations will bring our country to a new low. Even now, our country is the thirtieth in teaching in the world.

Down and down we go, where we stop, no one shall know.

# The Search For Peace

A place of quiet rest near to the heart of God.

Where in the world can you find rest? In a bar or any other evil place?

Can you receive a kind word from someone filled with hate?

Can you receive a word that takes your anxiety away when they are more anxious than you?

Will anyone invite you to their home when the door is closed?

Who shall bring a smile and a helpful hand when you have fallen down?

Who shall forgive me when I have done wrong in the judgment of many?

Who shall fight for my right to be free when I am in bondage to myself?

When I scream in the darkness of the night, who shall turn the lights on for me to see?

When I am losing my hope and faith, who shall restore them for me?

When I am screaming for forgiveness, who shall have mercy for me?

Who shall deliver me from all this pain? It is He who shall forgive my sins, and salvation shall befall.

# Are Aliens Here?

Are UFOs real and advanced flying objects of an advanced civilization?

The probability of an advanced civilization way beyond ours, observing our civilization, is very small. They would exceed our current total knowledge by thousands of years in the comprehension of reality.

Did you ever step upon an ant and not care at all?

The probability that mankind would be like ants to them is high. Our total civilization could be annihilated in a moment or two, just as we cause the extinction of the animals that inhabit our earth. Whether they are bipedal and carbon-based living entities is left for conjecture. No actual form of an alien has been shown to the world so far. Imagine a far-advanced alien reaching our solar system from a billion light-years away; that would be a phenomenal achievement. The probability that aliens are watching us is very small, but anything can happen, even if it's against the odds; that's all!!

# Cemeteries All Gone Everyone

Parents, grandparents, aunts, and uncles are gone to cemeteries buried everywhere. Brothers, sisters, and cousins have joined in their graves with everyone.

Many friends have joined in their graves that long did I know?

Girls once very pretty that I dated were now gone, lying in their final resting places, and many college friends joined them.

Where did all those young faces disappear to cemeteries everywhere?

Enjoy everybody whose young faces appear; for what seems a short time, they will disappear in lost cemeteries, everyone everywhere.

# One Year, You Are Young

One year, you are young, and as time passes, you become old. Each year can change your whole life. One year can bring you success, and the next failure. One year, you are single, and the next year you get married. One year, you are a senior in college, and the next, you get your degree. One year, you have no children, and the next year, you have your firstborn. One year, you are married, and the next, you get divorced, plus next year, alimony and child support. One year, you make an average salary, and the next, you are promoted and make more. One year, you are renting your house, and the next, you live on your own. One year, your children will live at home, and some years later, they will live on their own. One year, you are young, and as the fly passes rapidly, you have become old.

# Fun

Having fun is enjoying something special you like to do, like sailing on a boat, enjoying the ocean in the midday sun, or sipping a special drink.

Going to Las Vegas and playing your favorite slot machine or a good card game.

Have fun going to your favorite buffet and eating all the foods that you like.

Having fun going to another country to visit, like Greece, Rome, or Egypt, and seeing the pyramids or an ancient temple.

Having fun is attending a party with friends, laughing, and having something good to drink.

Having fun in playing your favorite sport, from bowling, tennis, or baseball.

Having fun in going swimming and sunbathing beneath the sun.

Having fun in many ways in the thrill of living life.

# Man A Million Years From Now

If one believed that man evolved from an advanced primate approximately three hundred thousand years ago, then what would he be like a million years from now?

Let's consider body hair first from all. Changes in our DNA, by probability, will completely eliminate it. Hair was upon our bodies to protect us from cold and rainy weather conditions in our beginning as other primates. A larger cranium with a larger brain is imminent, many times smarter than we are now. They are able to control the weather and almost all of our environment with instrumentation far advanced from today.

Life spans hundreds of years, free from disease, and with no body aches. Possibly smaller appendages in no physical labor we must, with enhanced sense modalities due to a larger brain.

Man will have a much greater perception of our reality. Mental telepathy is a great possibility with a larger brain; in other words, we will be able to read each other's thoughts. We won't be able to lie to each other anymore, possibly a much better society. Skin with fewer impurities in less exposure and clothes that never get dirty. There are no nails on the hands and feet they are becoming smaller.

Man will create all forms of energy devices doing all the work needed to be done in living in an almost perfect society. Transportation devices controlled by man's mind go almost anywhere in the universe. Smaller digestive systems are needed when eating special energy foods in smaller quantities. Breathing perfect air at all times with much smaller lungs. Reproduction is not needed as much with smaller sex organs, excretion possibly just a few times a month. Extinction of wild plant and animal life has gone a long time in a million years. It is not a perfect reality, but it is a thousand times better than ours. Take it or leave it, that's all.

# An Old Man Living By The Sea

There was an old man who lived by the sea, lived in a simple hut, and poor was he.

He had a little boat and started fishing before the sun came out. A young boy fished at his side as the waves moved in the morning tide.

The old man fished all of his life in good days and those of strife.

One day he caught the largest fish he had ever seen before, for all to see from shore to shore.

The young friend brought his morning coffee and paper to read; what a great fish caught, as everyone agreed.

# The Meaning Of Self-Esteem

The meaning of yourself defined from within is your self-esteem or your failure complex that predominates which choices you will make in creating the totality of your life.

A confident person in facing the various challenges will be calm and persevere over the tests that they shall endure. Those of low esteem shall be anxious in facing problems and fail before they begin the various tests they may endure.

Psychological maturity is very important in obtaining success or failure in the performance of which are considered most important to manifest life.

Those individuals becoming scientists, doctors, or lawyers must study for years to finish the course. Most who don't even take the time to see if they qualify have lost the ability to succeed.

Persistence is usually not a quality of those with anxiety and no peace of mind. Those who can read well at an early age will usually outperform those who usually have a hard time in reading and spelling.

If only parents with children with problems would get some professional help to avert failure, that could change everything. Contention after contention that isn't met shall knock you out in facing a successful life.

# Repetition

Repetition is a variation that must be in the learning process. Children repeat their multiplication tables over and over again, going from two times two to nine times nine until they can recite right every time.

Repetition in performing a task like tying the shoelaces of your shoe, and learning to use a fork and knife.

Repetition in using a toothbrush and in combing their hair. Increasing the complexity in spelling large words with the definitions known by just a few.

Repetitions in learning how to drive a car, and changing a tire when it's flat. Repetitions in learning how to read a poem, or writing something new.

Repetitions in learning how to hit a baseball, or in riding a two-wheel bicycle.

Repetitions in learning a college course for good grades to follow, and achieving a college degree.

Repeating the word success until it does occur. Obtaining self-assurance and confidence until it becomes part of your behavior in positive relationships in the needed repetitions to learn it all.

Repetitions are needed in all phases of a well-rounded life.

# Fire To Fire

Fire to fire as ashes lie upon the ground. Dogs howl in the disaster they see committed by mankind. Many words did occur before the bombs did fall, as the dead bird lies with the broken branches in the smoldering smoke.

A few were left to explain why the world became insane in a moment of regret that they could not undo the destruction everywhere.

The few that remain will take a hundred years or two to rebuild a world they destroyed. Choo-Choo watches for the destruction train coming very soon. So-long!

# Essay On Life's Desire

My life's desire was to consume ideas profoundly to prosper others and abound.

To consume my thoughts to create their own is the best to be found.

In the prosperity of new knowledge, a greater reality will be revealed for all to see.

Let's all get together and think in one accord for a better world to be free.

# All Life Must Prevail

All life must prevail and not suffer the extinction by man. The roaring of the lion, the king of the jungle, and the cries of the monkey where it does stand.

To enjoy the setting sun and the moonlight in the night, to hear the waves falling upon the shore, and to see the bird sitting in sight.

In seeing the great sea and the high mountaintops in view, maybe both only viewed by just a few.

Wake up from your dreams and ignore all the cruelty created by others every day; let no one escape the crime against other living things and make them pay and pay every day.

# Primates

Primates are the most advanced mammals in the world, including gorillas, chimpanzees, and orangutans, the smartest animals in the world.

Orangutans live on the island of Sumatra near southeastern Africa. They spend most of their lives in trees in the performance of their life cycle.

Orangutans are considered to be one of the two smartest animals of all. Chimpanzees live in the rainforests of Africa, living mostly in trees and on the ground. There are many species of monkeys, mainly those in Africa, South America, and India.

Those monkeys inhabiting Africa are called old-world monkeys, usually without tails. The monkeys of South America are called new-world monkeys, usually with tails. Lemurs are the lower primates in intelligence.

The gorillas are inhabitants of Africa, usually living in groups with a leader called a Silverback. Gorillas are considered one of the most mentally advanced primates of them all.

Most advanced primates can use types of tools to get their food. Next time you see an advanced primate, shake their hand; he may be smarter than you.

# Inflation

Inflation is causing the price of everything to rise, as I found out when I went to the restaurant to dine.

I ordered a steak and potatoes and my favorite dessert, apple pie. The waiter came back with my order, missing the dessert, which was just the steak and some applesauce.

I inquired about the reason for no dessert, and he responded, "Due to inflation, we can't afford to hire a bakery cook."

Received the bill which was higher than before. The waiter said, "you can add a tip over here," I responded, "can't add a tip due to inflation," as we left with an angry face.

Inflation, inflation, better ask for a raise at work!

# The Universe And The Merry-Go-Round

The universe is about fourteen billion years old, given a year or two expanding on and on. You may say, where is the universe going?

It doesn't know yet, but maybe it will be at a place to end or restart. Maybe the big-bang theory is just a farce; it just started by itself in nothingness.

Maybe you say it started from a universe that just collapsed; it could be so. We would need a theoretical physicist with the Webb telescope to tell us what really happened about fourteen billion years ago. But astronomers are too busy trying to define black matter and energy, if they exist at all. I am going around a circle of confusion if a star is really a star.

Many exoplanets are being found in atmospheres that abound; maybe one with a greater civilization than ours will be found. I doubt that idea—they would have found us first, and we would possibly be on their dinner course.

Let's get back to how the universe began; scientists need more facts and greater mathematical comprehension to create a new theory in maybe ten years or so. I shall go back to the merry-go-round; at least I knew that I was traveling in a circle.

# Clichés

An old cliché: what you don't know won't hurt you at all. What you do know may make it better after all. Maybe you take a pill for anxiety and avoid everything you know.

An old cliché: what's good for John should be good for Joe.

What do you mean by good?

It can mean anything you want. A good wife for John or being single for Joe.

A cliché: it's better to try than not at all. People have tried to go over Niagara Falls and have ended up in the hospital; he should have taken ten.

A cliché: better to fight than run away; the person I tried to fight was seven feet tall with muscles from head to foot. After one glance at that giant, I ran all the way home.

A cliché of old: better to have a penny than nothing at all. Today, a penny is nothing at all.

A cliché: go with an ugly woman who will be more faithful than a beautiful one. It's better to go with both.

A cliché a day will put anyone to sleep.

# Friends And Money

Take time to buy me a glass of expensive wine; you only have enough to buy a glass of beer. That means you haven't enough to pay me back the hundred dollars you borrowed last week.

With many more friends like you, I will need to take a loan from a bank to pay my bills.

You will pay me back when you win the lottery. I suppose you want to borrow the cost of buying those tickets also.

For any more friends like you, I must apply for welfare and food stamps in a life of poverty. A sad story with no end, that's all.

# To The Victor Go The Spoils

To Victor go the spoils mean the strongest gains the power and riches in defeating their opponent.

A whole country can be defeated and occupied by another and is ruled completely in every way.

This process of occupation has been done since civilization began. It seems that war is the inherent nature of mankind, and no solution has been given to stop this bloody process at all.

It seems that there lies within the unconscious mind the need to murder and kill with no restrictions at all.

One nation that is generally peaceful is attacked by an aggressive nation to victimize and kill and completely rule.

The mathematics of it all is in counting the dead in the brutality of war. Treaties, compromises, and concessions stop hostilities for some time and are broken to begin hostility again and again.

Bring your rifle and drop your bomb; kill as many as you can; what matters the most is who shall win.

# Merry-Go-Round Of Chemistry

As I went around the merry-go-round,

Each circumference brought me back to the point I started,

And this seemed meaningless in the time passed in living.

Dreams in the twilight fade from memory,

To be lost in the dawning of a new sunlit day of events yet to be determined.

The concept of daily chores shall be mundane,

And with certainty, they shall have little effect on social endeavors.

Compassion is a gift that can be shared with anyone interested in the golden light within it.

Delight in the moment that an atom without living thought can join its electrons in orbit with others,

To create molecules, forming the elements that are the foundation of all life.

In the sun's heat, hydrogen fused with molecules to yield helium,

To lithium, to create iron in final power.

In common sense, what should the title be in purposive tile science called chemistry?

In the model called the periodic chart,

All the elements known are listed in their complexity to form the reality that we perceive.

From radium discovered by Marie Curie to uranium in fission weapons,

Or explosive energy that may destroy the mind that created it.

The element called carbon emerged from the chart with a valence of plus one,

To be the foundation of the science of organic chemistry,

Forming the complex molecules in the miracle called life.

Amino acid molecules to protein chains to nucleic acids,

Giving birth to the vast arm of life that inhabits our planet called Earth.

The molecular configuration that exceeds them all is two hydrogen atoms and one oxygen atom,

Called H2O or simply called water.

Without this liquid, possibly no living form could exist.

This is Homo sapiens, the most highly evolved molecular configuration on Earth,

Creating the sciences that give him the instrumentation to unlock the secrets of the infinite universe.

Matter nor energy cannot be created or destroyed,

Yet all this discourse of knowledge can be lost to infinity on the extinction of man.

The sun shall become a nova and the Earth a cinder,

And who shall take note that it ever was at all?

# Meaningless Wisdom

Wise mouths enlighten the hearts of those who can hear their words of contentment and peace,

In abating the anger that lies within.

The good mind cannot be equated to an evil mind devoid of morality in the comprehension of it.

Judge, not the viper that shall not be dissuaded in unleashing its poisonous bite,

For death shall befall upon you in your condemnation.

A friend shall become an enemy in a moment of misconduct that challenges his vanity.

Do not despair in the lost moments of the past,

But rejoice in a single moment that brings you pleasure.

Do not dwell upon the failures of yesteryear,

But thrive on the success that has given you the ability to overcome them.

The manifestations of behavior depend on the individual's capabilities to interpret the stimuli and their conclusive response.

Learned social mores depend on one's ability to interpret the response correlated to their psychological maturity.

Those who seek to steal your identity shall next take your soul.

Spring blossoms the flowers in its sweet aroma,

And winter blows its cool breath on the leafless tree.

As all the deeds of man are performed on the rotating Earth,

Beware that all may come to an end in man's insanity that blackens the sky with his misdeeds.

Discoveries of the generations of man have given him greater comfort in the various aspects of his inventions,

Yet wisdom has not prevailed in the destructive events they have created.

Man, in all his creativity, is subject to the outcomes of the natural forces of his hostile habitat.

What is consumed by man in his need shall be replaced by another that shall be consumed tomorrow,

Yet this cycle of consumption must end with the lack of resources to replace them.

Dreams are manifestations of latent desires that cannot be performed at the stage of life.

Therefore, we are but a moral conflict that can only be resolved in the path of least resistance,

To satiate a mind that is cognizant of its corruptive state.

Fellowship with good people in the merriment of laughter and conversation,

For there shall be a time when the space inhabited by your generation is no more,

And the pictures of events are misplaced in the void of lost memories.

History is the recorded deeds of the generations of man,

Read by those desiring to learn of them,

Yet, those who performed them can read no more.

The countless treasures of antiquity are exhibited in museums,

For those to see the creativity of man,

But the ages will consume all things in our senseless mortality.

Probability determines the course of events in the future,

And certainty is the outcome of the known past.

The present is but a moment beneath the sunlit sky.

Man is but a resident in the confusion of the variables of reality,

And cannot evade the outcomes of existence.

Dark with vanity,

The humble shall be trodden underfoot,

And the vain shall inherit the Earth with the evil of its greed and power.

The contemptuous smile of the deceiver shall corrupt even those who contend to be wise,

And the victims shall tread the streets in their despair.

The sweet aroma of success of accomplished deeds shall erode to the foul smell of failure,

And the cycle of our endeavors shall hold us accountable for the finality of their outcomes.

Seek the mystery of the misery of reality in contemplation of the significance of its existence,

And the meaning shall be lost between two moments of time.

What is profound is manifested in the centuries the creative concept has endured,

To culminate in a millennium of profound concepts.

The essence of being is lost in the ignorance of the meaning of its construction,

In relation to the contemplation of the significance of reality.

The jester shall give laughter to those who watch his foolish antics,

Only to come upon the realization that they are laughing at themselves.

In a lifetime of dreams, we are but a commodity in the misfortunes that befall us,

In the foolish conception that we wished.

The games in the amusement of life must come to an end with applause to the victor,

Yet the victor shall suffer defeat to each mortality.

The stars twinkle in the night sky to enlighten man of his arrogance in solving the mystery of them.

The truth lies in the light, for it shall not be hidden from those who shall bear witness to it,

But a lie shall be found in the darkness destructive to those who believe it.

The wisdom of man is but a façade that shall lead him to a road leading nowhere,

And his knowledge is but a vain attempt to solve the mystery of creation.

Foolish mortal, you have created a world of confusion that shall lead to your demise,

But there shall be greater wisdom in the knowledge of the love of Him who has created all things.

# Inconsistencies

We are a moment of thought to be pondered and then lost to its meaningless conclusion.

Who can prevail in a continuum of probabilities that cannot foretell any moment of existence?

The blossoming flower has no definition when its petals have fallen on a cold winter day,

And the bird has left with its song of joy, leaving us in a state of sorrow.

The bee has taken its honey from our taste buds,

And the bitter taste of life ends in a woeful conclusion.

Whoever enters the gates of wanton desire shall tread a path of endless torture and pain,

And the crow shall be lost forevermore as it flies into the darkness of a black cloud.

The clock is an objective instrument for defining time,

As its hands are in a circular movement, it gives meaning to our existence.

Yet, what meaning does it have when its alarm sounds at the moment of our death?

Awaken from a dream in the darkness of the night only to the cognition that we were not asleep,

But lost in the infinity of dreams.

What has come from the womb shall be dust on a distant star.

Atoms join to become molecules that yield to neurons,

Yet yield scientific significance to reality in its elasticity as it expands into the darkness.

There is no morality in thought, for it has no barriers in its contemplation,

And its concept has no limitation in its creativity.

Wisdom lies in the most feasible probability that shall yield the most meaningful universal conclusion.

Run the streets of knowledge, seeking the portals to dimensions unknown,

And traverse time's infinite infinities.

Microbes, in a billion years of collective conscious probabilities, evolved into multicellular forms,

Enhancing environmental survival.

Evolution is not repetitive or linear in a given number of probabilities; it may be possible.

Man has tempted the fates and created the possibility of his extinction both knowingly and unknowingly.

His ignorance lies in the totality of generations that have created technologies that will master him or destroy him.

Desire has consumed your souls, and hatred has made enemies of each other.

Vanity shall give you no joy in living.

You shall be consumed by your own injustice,

And you shall have no peace in a mind filled with perversion.

Your purpose is an enigma lost in the carbon construct that gives one habitat of the psychophysiological processes,

Seeking a meaningful interpretation lost in a nonsensical vacuum.

Conscious observation yields unconscious destruction,

Enhanced by the immoral nature of our being.

Joy is a fleeting moment.

The light is vanquished by the darkness of our mental processes of self-destruction,

Our vulnerability is reflected in our image as it is fading in the mirror of contempt.

As countless precepts lead to concepts of creativity contemplating the infinite variables of our dimensional state of psychophysiological being,

Nothing can be assumed as time is constant in our certain extinction,

Which cannot be altered by countless generations of technological advancement.

The man was once subject to his sense modalities,

But with his ability to create instrumental devices,

He enabled himself to visualize the spectrum of the universe by studying its mysteries.

Homo sapiens, upon extinction, shall evolve to what species that could comprehend the dimensional state that is his infinite boundary in dimensions not known.

If the cerebrum of verbal and spatial conceptions is the totality of our endeavor to conceive of the enigma of extinction,

Then, what truth will be significant in the final result?

# A Plea Of A Sinner

There shall be a new world of love and joy eternal,

In the light of Almighty God.

Clothed in white gowns,

Adoring our forms in the incorruptible resurrection,

As the sons and daughters of God forevermore.

The light of God shall shine for eternity on all who believe,

In the sacrifice of love on the cross,

For all sin till the end of time,

By our Savior, Jesus Christ.

All is meaningless in man's own justification of his existence.

In all his creative ability, he is blinded in his vain attempt at his glorification.

In his attempt to master all of creation, he has become oblivious to the Master who has created him.

We are a pillar of salt in the disobedience of Almighty God,

For His ways are just and cannot be changed.

Do not lie in the darkness of fear,

But seek the light of the Lord,

That shall give peace and hope in a world that is chaotic and steeped in perversion.

Salvation lies at the cross in His sacrifice,

That His love shall lay in our retentive hearts.

So, the corruption that bound us shall fall,

And compassion shall dwell within.

The mind of logical conception shall avoid His spirit, which is unseen,

And seek the spectrum of reality that is seen.

Those who deceive others in their vain philosophy shall be judged by the content of its false doctrine.

Those who teach the truth in the Holy Scriptures find the path that leads to the kingdom of heaven.

God shall know the deeds of each man as he lived them.

We are blinded by the mask of reality that prevents us from seeing the only truth,

Which is our Savior Jesus.

Thanks to the Holy Spirit, who has revealed His truth.

To those of us who sought it, He opened our eyes to the light of His love.

Resolve in summation that the world shall be consumed by the evil,

That prevails in the minds of those who exalt in it.

In the end, those who have prevailed in the knowledge of God's righteousness shall never die,

But dwell in God's eternal love forevermore.

The good in the name of Christ shall prosper after you,

But the evil one has done shall consume his soul with the dust of his bones.

Whoever glorifies in his own vanity shall perish in his ego of self-destruction,

Yet those who have compassion for a brother in need shall enter into the kingdom of heaven.

Praise to the Lord who has created all things that are good,

And thanks to Him who sacrificed Himself for our salvation.

As an outcast and sinner in repentance,

Allow me to enter the gate to eternal life.

# The Hand Of Death

The faces in the dark walk to and fro,

Lost in their vain contemplations,

Seeking to survive in that brief moment called life.

The moment of youth fades in the passing of time into an aged body ravaged by decay.

Smile not on a sunlit day that shall pass to the darkness of sorrow,

For the hand of death shall silence mouths where the words are lost forevermore.

The laughter in living cannot be heard from the dust that lies in the coffins in graves laden with the saws of despair.

Death sits in the hidden place, resting for a brief moment,

Until his hand reaches forth in the touch eternal,

Which closes our eyes in the darkness that shall never see the light forevermore.

Horrible visions of spirits dancing in the fires of eternal damnation, never to escape the pain that burns their flesh,

For their screams in torturous agony shall never cease.

The mighty become weak in seeing their tortured souls in the reflections of the mirrors of sin,

And all is meaningless as they weep without end.

Honor not what we have spoken in times of glory,

For that time of victory shall be lost in the final breath of life.

Each generation shall bury their dead and put flowers on the graves of those lost souls wandering in eternal darkness.

Each smile renders a tear of dismay as laughter yields to the screams of discontent and agony.

As knowledge is gained in a lifetime of study,

So it shall be lost in a fleeting moment where darkness pervades forevermore.

Ancient stories have been written in the Book of the Dead,

For they shall never come forth from the tombs their bones inhabit,

And the sun shall pass over the pyramids for thousands of years in their final resting place.

Death has no preference over great kings crowned in gold or the very poor adorned in rags.

None shall escape the final sting of death as the candle wanes to its final light in its melted wax.

The season has come when the trees have shed their leaves,

Fallen to the Earth, and the branches are bare against the gray sky of death.

Nothing lasts forever as our dust is blown in the wind,

Lost in the clouds of despair, never to be seen again.

Memories are lost in the dreamless brain devoid of all thoughts,

And laughter shall be heard no more.

Fear is our continual companion in a world of gloom where tears fall in the sands of living.

Panic causes us to flee to streets of eternal doom in the horror of insanity,

And peace shall never be found.

All is an illusion. In the brief moment, you are a part of reality.

Death walks everywhere, plundering the essence of life and making all a shadow that fades into the darkness.

Brief light fades into the horizon, yielding us victims to the darkness of the night,

Wandering damned, lost in our sinful souls.

All shall perish in a world doomed from the beginning of time and a universe that will be lost in expanding space.

In the comprehension of our mortality, there is no conscious construct that will prevent that moment of our extinction.

In finality, man is a meaningless creature.

With all his intellect, he shall be food for the worms as his skeletal remains turn to dust.

Somewhere on a barren earth devoid of all life is a tombstone which reads:

Homo sapiens was born seeking wisdom and died in his own ignorance.

# Abstractions

The circumference of a lifespan equals the constant PI of life and death,

Times the radius of the events in the life cycle squared by the difficulty in its comprehension.

Consciousness cannot be comprehended on a holistic scientific analysis,

For it is an enigmatic puzzle that is eternal to neurochemical transmission.

Abstractions surpass the conception of reality to the insights that enhance the ability to creativity.

This creativity may be intangible to analytical reality and its conceptions,

And may bridge to a greater infinite design.

Abstracting reality has no limitations in transcending the concepts that define our dimensional state.

Abstractions enhance creativity to conceptions not yet contemplated.

Preconceptions evolve into conceptions in multiplicity to the abstractions of greater comprehension.

A stimulus evolves a response that is appropriate to their relationship,

Which is fundamental behaviorism.

The complexity of the above phenomena depends on brain development.

Abstractions as a thought process surpass behavioral reaction to reality,

Even to its highest level.

If reality were abstracted, it would go beyond unmeasured infinity.

There is a parallelism between abstraction and creativity,

Yet, they cannot be transversed into a conclusive comprehension.

What is theorized and its conclusions are deductive probes into the comprehension of reality,

However, the abstracted theorem creates a realism that may not be real in known dementia comprehension.

We are mere mortals walking in the shadow of our ignorance,

Yet, abstracting the shadow beyond that formed gives us an insight into a reality that is not yet known.

Abstract The moon and the craters shall yield the asteroids that formed them.

Abstract the billions of galaxies that are the inhabitants of our universe,

And you shall go beyond singularity into the darkness of creativity.

Abstract all the twinkling stars in the evening sky,

And you shall become the energy that sustains them.

Wisdom is an abstraction of the mind that gives greater meaning to truth,

For truth is the shortest distance to the meaningful comprehension of ourselves.

An abstraction goes beyond the speed of light to infinite universes not yet formed in a moment of singularity.

The raindrops are abstracted into the red rose that blossoms to carve spring,

And the snowflakes that fall to the Earth and color the field white.

The violin strings are abstracted into the notes that give the listener musical compositions to enlighten his mind.

The dreamer has his dream, and the mathematician has his equation,

The abstraction is the dream that equals the equation.

Painter, put your brush down, for your painting is an abstraction with no meaning except to the beholder.

Are we but a fragment of one's imagination,

To be abstracted in their fantasies in the darkness of nowhere,

Or are we the reality of the insanity of nonsensical words that are uttered in meaningless prattle?

In the serenity of the womb in labored expulsion,

We are thrown into a world of unrest

Only to be abstracted into a utopia of peace in the brightness of the light of unknown origin.

Wonders of wonders, the lion has ingested his last meal and shall roar no more.

In these conclusions, abstractions can make what is meaningless meaningful,

And this thought process can fill the chalice with the wine of wisdom to comprehend the origin of our being.

Abstractions conclude in finality what has not been concluded in all man's endeavors to contemplate his bipedal walk on the paths of infinite discoveries.

Abstractions embellish my mind with all the probabilities in a continuum of events that shall evolve in combinations that shall traverse infinity.

# The Asteroids

Long-lasting waves of thought traveled the universe,

Passing a lost asteroid traveling in space with a course uncharted to a destination unknown.

The asteroids, once lit by the sunlight of a dead sun,

Are passing in the darkness in a sleep that is unending.

What is in its crust are lost centuries that have no names on the tombstones,

Barely visible to the eye of an unknown observer.

Memories are turned to dust,

And laughter has been silenced by the faded apparitions that disappeared forevermore.

The songs of nature that were once sung on your crust are but a requiem of doom,

That could only be heard by the distant stars that surround your moving mass.

The birds have fallen from the dead trees that once stood in their glory but are no longer standing at all,

The blue sky that once seemed to have no depth has lost its hue to a dying sun.

The shadows of those who moved in the many years of living are left alone to move amongst the many asteroids,

That mourn in the silence of space.

Tongues that once uttered words of wisdom have been lost to the decayed flesh that gave them the ability to speak.

The damned are even more damned,

And the cursed are even more cursed,

For they are not even cognizant of their eternal flight.

The dust on the surface is all that remains of the giant edifices that reached heaven above.

Stars glimmer in coded reference to the unique solar system that this lost Asteroid was a member of.

Gamma bursts applaud the beauty it displayed in the light of the friendly sun.

Even the galaxy marveled at this speck as one of its favored inhabitants.

Lost Asteroid, even the clouds that once surrounded your surface would shed raindrops as tears for your lost glory,

And the thunder sounds its anger at your fate.

Love that was shared has been lost on your dark surface,

And the warm hands entwined have become cold forever more.

The butterfly shall never come from its cocoon,

And the flower's array of colors shall never be seen again.

And the rainbow shall never arc across the sky.

Little Asteroid, you shall never again be clothed in such a wondrous display.

Even the trenches of war are no longer seen in this strange species' anger of warfare,

And the lingering sorrow of death that follows.

The violin and the piano played as one,

And so the notes cannot be heard in music that enlightened those who heard them.

Comets belittle you, and stardust cannot give you rebirth in the drama of infinite change.

For time steals the beauty of the rose of springtime and causes it to wither.

Your companion called the moon has become debris in space and shall never again display its moonbeams in the evening sky.

Little Asteroid, where are all the dreams that were dreamt about you on a warm summer night,

And the technology that gave your surface the power to see the universe and theorize its origin?

The mathematicians and their equations are no longer significant in the darkness of infinity,

And science has no meaning without the brain that sought knowledge in it.

Beware, Asteroid, that there is no other asteroid in your path that will collide with you in the void of outer space.

For there shall be countless rocks of debris in the aftermath of the final dust of extinction.

Walk in the finality of what was such a splendorous beginning as these pebbles are engulfed in the darkness of expanding space.

All is meaningless in the darkness of despair where all significance has ceased.

At a moment of the greatest improbability, there appeared a bright light shining on these rocks lost in infinity,

And they began to adhere to one another.

And behold a planet emerged that was sustained by the Holy light that shall never be extinguished.

The lost graves yield those that have now eternal names in a new heaven and Earth that will have eternal life in the glory of Almighty God!

# A Place To Nowhere

A time for life began in my early youth,

A time to learn the difference between a lie and the truth.

Can anyone separate imagination from the real thing?

What is real to believe, or is it just found in a nightly dream?

Deception can occur to those with little knowledge and not being,

Seeking a straight path that will lead to His light.

Uncertain behavior in seeking to solve what is unknown,

Creative new AI in dangerous drones and what to clone.

A trip to the moon and, in a few years, to the planet Mars,

Much greater technology is needed to reach the closest stars.

In this generation, phenomena discovered that are not known,

Going from the micro to the macro, and there are many new frontiers to explore.

Since 1995, generations have seen the sun shining in the sky.

Nowhere can the greater creation be found when Christ is the resurrection and life,

And thousands never die.

There is a love that dwells within,

Creating a smile of caring more than has ever been.

# Lost

What is lost is hard to replace. What is gained is even harder to sustain.

Hatred vanquished among each other, peace should be maintained.

Funds lost in badly made investments are hard to regain.

Lost time in pursuit of life's mysteries cannot be reclaimed.

What is lost and found is a thing to celebrate and applaud.

Nobody can predict future events, whether good or bad.

Sleep eludes me in thoughts of tomorrow's expected rain,

A moment of rest or peace or my mind is driven to the brink of insanity.

Pondering in an aged frame upon what was missed, not upon what was done,

Asking for forgiveness in His shed blood, eternal life was won.

Lost souls abandoned to the fires in hell below,

Not even a drop of water in a place forever to be.

The loser is a state of position not accepted in any society,

Hard to find a successful solution in your final story.

Seek first the kingdom of God; then all things will be given to you.

Walk through the open door to life eternal, always loving, and always true.

# Reflection

Reflections upon the endeavors of reality unknown,

Hypotheses and theories accepted are factually uncertain,

Creating insights into phenomena not seen.

To travel to the stars far beyond our capabilities is only a dream.

Let not our debris be the testament of our industrial might,

Sound thought and action can defer possible negative outcomes.

A lion in Africa poised to pounce on its prey mirrors our careful planning in every way.

Raising a family embellished with wisdom and love shall cause them to grow in self-esteem and success after all.

In failure and poverty, tears shall abide, and students in colleges seek knowledge while others face ignorance to befall.

Some earn good notices, professions, and degrees, while others endure menial work and behavior.

Distant stars await our feet, diseases to cure, all possibilities go beyond a dream.

Minds erecting new technologies increase our probability of being or not being.

For not to be is reality's grim plight, believe in Christ's spirituality,

Never to die under his eternal light.

# Paradox

Seeking purpose appears trivial compared to the stars above.

Today is the time to give the most for tomorrow's promise, not just to try.

Positioned in the gravity of a black hole, becoming a mere speck.

Cruelty towards loved ones creates singular fools.

Heads or tails, I have lost not a cent,

Success is equated to positive outcomes, not negativity.

Boys have soldiers, girls have dolls, and a struggle between masculinity and femininity.

What's this conflict all about? Hypothecate, expecting a result, computing hard to find a valid result.

A circle rotates better than a square, improved design, hypothesis to theorem, a new chain of reasoning, newly found.

A new truth of creativity enhances reality more profoundly.

Found accidentally crucial diseases in experiments thought wrong,

In the sum of humanity, living a healthy and long life and embracing nature, sadness is circled with darkness of despair.

Open the door forgotten, leading to hope and newly found cheer.

# A Smile

A smile, perhaps leading to laughter in a humorous moment,

It is best to happen when one does not know what is next.

A smile when dreams of success have come true.

A time of merriment to celebrate the good that happened for you.

A smile is like a gift that comes in a moment,

Likened to the rainbow that happens after the rain.

A smile comes from the heart when a blessing to the stars is given from above.

It is a miracle that fills your soul with perfect love.

A smile when your first child is born. That is a wonderful day.

It is a time to give thanks to a greater Father and pray.

A smile in times of trials and tribulations of bitter streaks.

For the light has vanquished the darkness to a greater vision to know.

A smile of friendship to those we once did not care for,

Giving moments of forgiveness and friendship happens everywhere.

A smile is helping those who have fallen and cannot rise on their own feet.

They shall enter the small gate to walk on that eternal street.

# Why

Tell me why the sky has a hue that is so blue,

As a vehicle for the birds and clouds to travel through.

Tell me why the sun gives light to enlighten our days,

It is a time to have a picnic and for the children to play.

Tell me why some live in anger and others forgive,

When some like only to receive, while others take joy in giving.

Tell me why there are many wars in a time to kill,

With free men's protests not heard and voices that are still.

Tell me why those who prosper give to those who need,

And those who keep all that they earn are not satisfied with their greed.

Tell me why the bird builds its nest in the safety of the tree,

Likened to our Savior whose salvation and love shall always be.

Tell me why I cannot see the wind blowing through the tree.

It is like our spirit that becomes suddenly free.

Tell me why you are the Word, knowing the end and the beginning,

Shedding your blood for all that have been sinning.

# No Other Love

No other love waves in fighting for the country in a time of war,

As heroes come forth in battles, their valor soaring.

Battles in defeat and victory, whether wrong or right,

See those from towns and cities trained to fight.

Is it our nature to settle disputes with many lives lost,

Loved ones to mourn, nations parading their destructive weapons,

And are militants showing much willingness to destroy others?

Sound the battle stations, weapons in hand, facing the fight to come.

Intensity increases, gun against gun. Will a victor finally emerge?

Debris of ships floating on the waves ships sinking below,

The enemy's ships are in full retreat as victory is seen with ships blowing their horns.

Sweethearts are divided from each other, enduring valiantly until peace is won.

Coming home from shore to embrace, crying for the casualties,

Never coming home to see the sky above.

Flags waving for the heroes marching in a country they love.

# Big Bang

If the event of the Big Bang evolved out of nothing, then the problem of another is to infinity in an absolute void. The decline of all forces in the universe must occur based on the energy expended in the Big Bang. No hypothesis is concrete on how the universe will end or partially end in current mathematical or more comprehensive insight at the present time. Man is in the primitive state of knowledge in discovering a civilization greater than ours that has existed or exists concurrently at this time. Our civilization may become extinct by many probable events, all of which are superfluous in our meaning. Our ability to think leads us to confront the stars, for on this small planet, we are meaningless to the time of infinity.

# Ideation

In deep contemplation,

The reflection of our imagery is founded on the conceptions that reason through variables,

Which culminated in a single moment of thought.

In that singular moment of thought,

Creative actions function to enlighten the mind with newly found insights into reality.

A single moment of ideation shall exist beyond the expanded universe,

Serving as the precursor to the formation of energy that causes the event of singularity.

Ideation continues long after the biological entity that created it has become extinct.

A parallelism exists between ideation and the generation of energy,

Generated by the probability of its occurrence.

Following this singularity is the entire phenomenon that will occur,

By the law of probability in that time for events to be consummated.

Once all outcomes in a single universe have concluded,

Then, a single ideation shall remain until the probability of singularity is reached.

The purpose of biological creativity lies in the meaningful consummation of ideation.

Reasoning must reveal that all events and outcomes occurring in the universe,

At any given moment, they are consummated by probable causes in the phenomenon of reality.

This parallels infinite realities that have outcomes dependent on the light-years of expansion within any given universe.

All outcomes of events, depending on their complexity, must come to an end,

Thereby making infinity infinite in the dimension of time.

The remaining ideation shall act as the catalyst for recreating countless singularities.

All forces, both known and unknown, that operate in the expanding universe of matter and energy shall decrease in intensity over time.

Atrophy and expansion of the universe coexist until the probability of any function negates another,

Leaving no defining factor of universal reality.

The function of any biological species capable of an analytical approach,

It is to comprehend their environment to perpetuate their existence with greater knowledge.

Extinction of any species will be probable,

Based on the assumption that the variable of extinction cannot be comprehended.

We are mounds of dust in a vain attempt to comprehend the natural phenomenon of reality and its continual variable changes.

Our biological construct remains a mystery until death claims all of its ignorant contemplations.

Man is but a primitive enterprise residing in a universe of phenomena too vast for him to fully understand.

From fire to the light bulb, he has progressed,

But does he see the light of life within himself?

Yet he has transitioned from fearing darkness to illuminating rooms in the night.

So, few properly serve the dish of knowledge and discovery to a vast populace ignorant of it,

Making civilization dependent on the few who have the creativity to perpetuate it.

Exploration is an inherent drive of our species,

And we shall continue advancing technological achievements;

What is impossible in one era may become probable in future eras to come.

The universe is but one place in the infinite wonders of reality.

Ideation is a continual process of creativity that surpasses any realities hypothesized by the mental conceptions of man.

In the fusion and fission of man's indulgence in creating weaponry,

Which may ultimately lead to his extinction,

His discoveries personify his ignorance of their results.

From hypothesis to theorem in deductive reasoning,

Man attempts to incorporate results as constructs of reality in his endeavors to increase knowledge.

In the variations of resultant deductive reasoning,

Lies the enigma in the design of reality that may be unraveled.

In summation,

The ideations of all advanced biological entities that inhabit universes,

Transcending reality in the consummation of all probable events,

Remain as the precursor to energy that creates potential big bangs in new universes of countless realities.

We are but tools bridging the finite to infinity in the creation of endless realities through the simplicity of ideation.

## Confusion

Delight in the moment of the day,

For the night shall consume it,

And the memory of it shall fade in the moments yet to come.

We are clothed in the process of our reasoning,

And become naked in the midnight hour of our dreams.

The secret of a smile shall remain a mystery,

Until spoken words reveal the joy in it,

And a shed tear shall fall in the mystery of its grief.

The righteous are pretentious,

And the good they do shall be lost in the dust of their deeds.

We are resolved in the daylight,

To compete in daily conflicts of our travails,

To complete the work in the field of our station.

The nightfall is a time of reprieve,

To ponder the results of our endeavors.

As the baby clings to the skirt of his mother's security and protection,

So does the adult cling to the knowledge he has attained,

In the measure of his success.

The result is culminated in the victories and defeats in living our life cycle.

The final result is written on a tombstone where life has ended, and it states, “Rest in peace.”

Vanity shall rival compassion in the course of human interaction,

The result is determined by which has the greatest impact on those who must bear witness to each.

Those who have fallen shall arise with those with compassion,

And those in sorrow shall be avoided by those of vanity.

The lust for life in the artistic purpose of regenerative activity,

It is dependent on moral levels of individual choice.

The result can be either creative or destructive.

Imagery is a continual interpretation of the reality that is our habitat.

Our thoughts are not reversed in the censured words of communication,

Therefore, we are actors in a play called Life.

What is likened to one is unlikened to another in deeper contemplation,

And honesty, in the final analysis, is a mere word.

The duality of thought and communicated verbiage,

It is a variable of human behavior,

And is partially understood in the motivation of behavior.

Therefore, man lives in a civilization of confusion,

Leading to encounters as critical as warfare in all its aspects.

A singular natural event can alter all hypotheses into the dust of nonsense.

Open the envelope to read the variables of life's finite meaning,

And the pages are blank as the conclusions have faded into infinity.

What transpires in a million years shall be lost in the next million years of time.

If the world were flat, then the solar system would be a rectangle,

In a universe shaped as a cube in the formulation of our ignorance.

The errors of continuous conclusions shall reveal our inability to interplay with reality;

We are but primitive forms wandering in the limited space of our ignorance.

Thirst shall drive us to sell our souls to avoid the hand of death,

For we are only biological entities seeking to survive a moment in the conception of our finality.

We have reaped a lifetime of deeds in a body aged with long-lost conceptions.

A jest of yesteryear only transcends to a word of hopeless contempt,

And accomplished deeds in the victories of life have succumbed in years of defeats.

The past in time shall consume the present to a future of unpredictable events.

Thought shall delineate our vanity,

And pride shall vanish in a chair called weary bones.

Pictures at an exhibition can depict a life of confusion embellished with uncertainties,

In the outcomes of events in the attempt to live it.

The light of one moment becomes the darkness of another,

And then we escape to sleep in a moment of nonsense.

Some have faith in the spiritual unity with a supreme being,

Believing that its reward shall vanquish the sting of death in an eternal state in the afterlife.

Others are devoid of any spiritual faith,

Loving only their reflection in a mirror of lost hope and sin.

The inventions of man across generations have expanded his knowledge of reality,

And his ability to live a wiser and longer life.

But in his ignorance, he may invent tools of his extinction,

Just to see the light of love that lies hidden from hearts of stone for a moment of peace that shall never be forgotten.

Weary traveler, your road continues until you have fallen,

And your journey ends in an inn of eternal residence in a room with no bed.

You have spoken to many who were wise,

Yet what you have sought you could not find,

Which is the greatest: love.

Walk through the green field filled with the aroma of blossoming flowers,

And the light in the blue sky above for a brief time,

For the cold hand of winter shall expunge the fruits of springtime,

And leave empty fields under a gray-cast sky.

Be concerned but for a brief moment,

For the fruits of heaven shall spring eternal.

Where were all the people present in pleasant moments of joyful friendship?

Consumed in the passage of time, all of them.

I stand in contempt, blaming the moonlit sky for betraying me in the darkness of the night.

Do not delay in calling sleep to a place of peaceful contentment,

Where the shepherd shall watch over them as the ravishing wolves are held at bay,

And they shall graze on the food of everlasting life.

We stand in confusion as the edifices of our cities fall upon us and crush many in their haughty display;

They worshipped themselves in the idolatry of their accomplishments,

And did not honor the one who created them.

The end of life is in the confusion of the beginning,

And all our deeds lie fruitless in the mystery of our mortality.

# Peace

Those who thought they were wise have been condemned fully from the time of creation. Wisdom is a gift from God and not the boastful conception of that professional knowledge.

The tree reaches to the sky, and in the sky, it reaches to God, who created all things. Only the foolish cannot see God's words in the creation, for the darkness has blinded their eyes and consumed their soul. Those who believe they have not sinned have no knowledge of the mental conceptions they have derived, for they have been overtaken by the happy existence that sustains them. As the sea rushes to the shore to be lost in the sand, so does their concept become void in seeing the truth.

In sorrow, there is a victory in that joy has come from above in the knowledge of his eternal truth. In seasons, the tree loses its garb only to be clothed once again, and the flower is lost in the snowflakes that reappear in all their glory in the bright sunlight.

# Facing The Many Facets Of Life

When various problems in life create too much stress, take some time to think about the best answer for each one of you. Be pragmatic, not idealistic, when solving problems that confront your mind, and often share to see if your conclusions are proper for you.

Each new day, in its special way, will be a new challenge to face.

Some days are nice and sunny, other days with dark clouds and overcast. Problems are part of life, and all must rise to solve them in the best way possible. Your reason for living is to balance life in your favor.

Your mind must be trained mentally and emotionally to face anything that may confront you.

Going to any school to get good grades and graduate takes a great effort to do. Getting married is probably one of the greatest events that will happen to you, and staying married is even harder to do.

Having children and raising them right is a great challenge to you.

Getting a professional license to make your family secure is a very wise thing to do.

Getting all the proper insurance to safeguard the many facets of your life is best for you.

Good luck! Good luck! It's a miracle if you survive it all.

# The Harvest Of Our Thoughts

The harvest of our thoughts shall create greater knowledge in a world that is not fully known.

Some insects standing ten feet or more three hundred million years ago became so small.

How did man's brain become so large that he rules the entire Earth above all? How astronomers believe that the sun will last over a billion years until it evolves so that no life will exist at all. How our solar system became eight planets, the first four solid and small, and the last gaseous and large.

How many animals are herbivores and others are carnivores as many species evolve and become extinct?

How the Earth spins around and orbits around the sun? How did life originate, and how do all living entities need water to exist? Why did the universe suddenly appear and seem to be going nowhere?

What is the future of mankind billions of years from now?

There are too many questions to answer about what man doesn't know; it is better to walk into the darkness of the unknown.

# Reality Is Not Known.

Reality is not known. No conclusive purpose for reality except that it's a constant and is a product of the present. Reality is as variable in the time it equates its identity. Only probability may predict its outcomes. Each step may be highly predictable, but one fall wasn't likely to occur. As time passes quickly, one outcome becomes harder to predict. It would be very hard to predict an event and outcome two days from now. The events and outcomes in the totality of reality surrounding you make it hard to describe the future at all. Anything can happen to you in a moment not predicted to befall.

# Hope Lies In Anyone

Hope lies in anyone who will take the time to solve your problems. Some problems require more thought to solve than ever before. Like when you don't have the money to pay your mortgage for this month or your car payment either. You don't have the credit to borrow any money or get a credit card. A relative helped you pay the mortgage last month. Most friends have the same problems as you in paying their bills. You lie awake, trying to figure out what to do. You could sell the house, which doesn't have the equity to pay all the bills. You could give the car back, but the creditor will sue to get the money back. You're already working two jobs a week to pay all the debt, and your wife is disabled. Going bankrupt will result in living on the street. What shall I do? I didn't tell you I would solve your problems at all.

# What Composes A Symphonic Orchestra

The composition of a symphonic orchestra is simplistic but amazing in its presentation. A great composer uses notes written in various scales to create the sounds of his symphony. There are various types of musical performances called sonatas, concertos, and special presentations for specific instruments. The symphonic orchestra is composed of various sections of instruments to present this great music. The first section can be called the strings, which are basically the violins, violas, cellos, and bass. Variations in the number of string instruments depend on the compositions. The next section, in no particular order, is called the horns, like the trumpets. Their volume can increase the intensity of the music played. The next major section of the orchestra is called the woodwinds, with a reed in the instruments. The clarinet is a good example of a woodwind instrument. The next section of the orchestra is the percussion section, mainly the drums. Harps can be added to the strings, as well as bells to the percussion instruments. A great conductor leads the orchestra in his presentation of a great symphony. Each section of the orchestra is played by great musicians to present the greatest music in the world. Presenting the works of Mozart, Haydn, and Schubert stimulates the minds of the intellectual.

# The Dumb And Dumber

It's dumb and dumber instead of smart and smarter. How far do we have to go before society drowns in its own stupidity? To perform the impossible is left to a few, while mediocrity entertains the most. The learned or the average, going forward instead of backward, our future lies in what we choose, after all.

# The Future

The future, the future, is waiting for anyone. The future comes and becomes the present, and the present becomes the past. All happens as quickly as the twinkling of an eye, not remembering much of the time passing so fast. Must accomplish this today, and I forget yesterday when tomorrow had already gone by. If you know what I mean, that's a big surprise. Tomorrow will become today, and today will become yesterday. Is that pretty clear? In watching my favorite football game tomorrow, I will know who won yesterday and forget today. Is that possible? Maybe yesterday, today, and tomorrow are the same, and time doesn't matter at all. Maybe I missed work yesterday and took the day off today. Who shall remember when my wages will befall? I ate supper before breakfast, breakfast before lunch, and a snack before them all. I locked the door to my house yesterday and plan to come back tomorrow. How can I open it today when I am gone? A prisoner asked the warden, "Did I finish serving my time for my crime?" The warden answered, "When did I admit you in?"

## Seeking The Wrong

To prefer negativity and defeat in living rather than a positive attitude and seek to win. Be sure what you choose in life is right for you. To smother everything rather than to seek to unfold. To seek what is bad rather than know what is good. To cause those to fall instead of lifting one up. To seek to put those into bondage instead of letting them be free. To seek to lie to everyone instead of telling the truth. To cause those to find sadness instead of joy. To create destruction everywhere you go instead of building everything up. To create poverty and hunger in many rather than prosperity for all.

# You Should Not Wish It

Showing you why you shouldn't wish for anything at all. You get three wishes from a genie or an animal's paw in wishing for anything at all. The first wish is for a large amount of money, which is granted and does befall. The money from the first wish is lost to taxes or a bad investment, to lose it all. The second wish is to become president or king of a country and have all the riches that come with it. The person isn't able to maintain his presidency or loses his kingdom, having no title at all. The third wish is usually wasted on getting mixed up in wishing for something small. Better not wish at all and work for what you get. Smarter, that's all. When you wish upon a star in this supposed magical world, stop; it's much better to pray than any magic at all. Disney has been deceiving us into thinking magic and wishing are beautiful to enthrall when they shouldn't be practiced by anyone, seeing the evil it teaches everybody to know. Disney teaches us that fairytales will create a good life instead of God, who knows exactly what we need from life's beginning to its end, after all. See how deceiving life can become in choosing evil over good. Beware, the light is really the darkness, after all.

# Harris And America Will Fall

Believe the writings are written upon the wall: if Harris wins, America shall fall. Her comprehension isn't genuine in defending America against Russia and China; she lacks the strength and knowledge to make America strong again. Extreme liberalism is biased and can't create a victorious people to rule with respect as number one in the world. Politicians must be strong and honest with the people to reduce hate and prejudice among all. America should put education first, learning about everything so we do not fall. We must carry the torch of victory and not back down in defending our freedom and liberty to anyone at all. May God bless us in everything that we do, after all.

# Mankind Problems

Accelerate your thoughts to places you could never think you could reach when discussing phenomena in depth. Mankind's dilemma is that it will take a multitude of different constructs to resolve its problems in relationship to reality. How do nations avert a third world war in opposing political positions with tensions not abated to any degree? Extreme poverty in many nations causes hunger and disease, with many not caring at all. Concurrent wars are dangerously being waged and are not popular in a divided world. Political vice is being practiced by most countries in their evil of power and greed. Nuclear weapons are pointed at each other until a nervous hand destroys countless cities to oblivion. This is the problem: nobody knows what the hell to do.

# Opportunity Knocks, But Once

Opportunity knocks only once, so how can opportunity knock at all? That money can grow upon a tree if you find that imaginary seed in reality at all. Just wish, and they shall come true; if you believe that, maybe you drank too much alcohol. Just wish upon a star; maybe you just got lost in the dark. Something is better than nothing at all, not if something broke your leg. Better to be quiet and not speak, not in speaking the truth. What you don't see can't hurt you— not true when someone is pointing a gun at you. Sticks and bones can hurt, but not words; false words can put you in jail. Let a smile be your umbrella; that doesn't stop you from getting wet. Any kind of job will do, but what if you're a night watchman in a cemetery? Better to have peace than to have war; explain that to the dead in our current wars today. Better quit while I am ahead!

# Man Vs Woman

Are women meant to dominate men in every aspect of life? In the Bible, Yahuoh created women as men's helpmates after all. Does that mean women should be slaves to men? No, not at all. Men are generally stronger and can perform jobs that require the tilling of the land that Yahuoh meant him to do. Women were to bear the children and prepare the needs of the family. In the beginning, that was the way it was going to be. Women of today should go to school and strive to obtain any degree they can in their education. In my generation, about eighty percent of professional schools were men, and twenty percent were women; now, women dominate at fifty-five percent. Men have lost their status in the performance of life and hide in getting a tattoo and another beer. That's one reason why sixty percent of American families are one paycheck away from the streets of poverty. I think masculinity and femininity have a purpose in being different in creation by Yahuoh. Man can lift two hundred fifty pounds, and woman one hundred seventy-five, and normally that will not change. Testosterone is made for war in the field, and estrogen generally is not. According to Almighty Yahuoh, men shall dominate once again until the end of the world, as He intended it to be. In the eternal city of Jerusalem, there will be no male and female in the flesh again. Yahuoh never changes in the creations He meant to be.

# Mickey Mouse

Walt Disney first presented a successful cartoon of Mickey Mouse in 1928 called Steamboat Willie in black and white. One year before the Great Depression, Walt produced more than twenty cartoons of Mickey Mouse that year. His first color cartoon of Mickey Mouse appeared in 1935 with great success, and a famous Mouse was presented around the world. How can anyone count the number of children that watched his cartoons of this famous Mouse? Next came Donald Duck, whom you could barely understand when he talked in the cartoon. Not to mention Pluto, the dumb dog. Walt was so successful in these cartoons and movies that he established the Walt Disney Studios. After months of consultations and evaluations, Disney started Disneyland in 1954 in southern California, opening a park in Florida called Disney World and other parks in other countries. Generations of countless children and adults enjoy the results of a single Mouse—one of the greatest concepts ever done. Now, if their giant corporation stays out of controversial issues where they don't belong and sticks to genuine entertainment, they might survive. Keep Mickey Mouse alive and mind your own business. So long!

# Sabbath On Saturday

Let Thy peace fall upon my human frame in knowing all that must be done. In Thy perfect time, all things shall befall where they need to be above all. Great is Thy Word that created the heavens and Earth in six days, and on the seventh day, You rested in Your perfect work. All knees You created shall bow before Thee no matter what judgment they shall receive. Let my faith grow and grow where it should be. For the Sabbath is on the seventh day, Saturday and that shall always be. Amen!

# What Makes Sense Today. Not Tomorrow

What makes sense today becomes ridiculous tomorrow, for wisdom has a great expanse in advancing knowledge fast. One archaeological find can change our perception of an ancient civilization in all that was previously known. One mathematical equation can change what is known about time and space after all. One chemical component added to a known drug may cure a type of malignant tumor. A new type of propulsion may cause rockets to travel millions of miles faster. Images created by our minds and projected upon screens offer a whole new product.

# Money Tree

I shall plant a money tree to provide me with the money I must have to succeed in everything I plan to do. I realize that each step toward success can be measured in the dollars needed, for nothing is free. Just give me an extra breath while waiting for that money to grow so I can pick the dollars off its branch. I shall use miracle water, blessed, to make sure the tree will grow and not rely upon chance. A miracle tree appears to me, for I need the dough.

# Sing And Dance

Let's sing along and dance a step or two to know where we belong in harmony as one for a few hours of joy. The rest of the time shall be spent wrestling with the problems surely to come. Spend some time running through the park, saying hello to those you pass by in the beautiful sunlight. Sing with the bird and buzz with the bee in those moments hard to find. Smile at the children playing their games and the old sitting on the bench as you pass quickly by. Smiles don't come easily; you must take time to capture a smile at least once each day. Take time to wink at the prettiest girl at the party. Above all, maybe she will stop and say hello. Jump over a high place and land anywhere you shall enjoy. I painted a pretty picture of what life can become, and I hope you join me in having a little fun for a moment or two. I hope we shall abide in the playground long lost in time but still in the lightness of our mind.

# Is Man An Animal Or Of God

The best expression is a smile when you intend to complete a meaningful thought. A meaningful thought has the possibility of discovering anything not known before. The probability that mankind obtained supremacy over all living species to rule the world may be smaller than you think. Those who hold to progressive evolution look to cranial expansion as a measurement of expressing intelligence. Our species is named Homo sapiens in scientific lore. The greatest question of all is: is man an animal or a creation of God? Science does not know how our consciousness works and how life began, and many questions of behavior remain a mystery. Science can see the sun and energies that abound but doesn't know the reason why it should be there at all. Man has created a great civilization that rules the Earth but has the ability to destroy it in a moment of ignorance. Does deep in the unconscious mind lie the spirit that The Almighty breathed in with a living soul to give us life above all? You choose: Homo sapiens or created in the image of God, dust forever, or eternal life?

# Consistent Emotional Stability

It's better in a state of abnormality to let the stress be properly expressed to reach a state of normality. A state of normality allows you to think more clearly about how hard this peaceful state was earned. Fluctuations in emotional expressions allow a person to reach a higher level of thought. When your emotions are out of control, your mind, in all its energy, seeks a better balance of expressions to occur like a utopia, creating a better state of thought in facing the facets of human strife. If you had all the money you need and peace of mind, you would ponder upon your mortality. Consistent emotional stability can't be achieved in the constructs of our mind. One moment, we are ready to ignite; the next, to hide in fright; and the next, to take a drink to forget it all. I think I shall sleep under my bed, hiding from tomorrow.

# A Moment Of This Earth

I am but a drop of water in the endless sea, a small wave abounding on the endless coast. I disappear between the endless stars, just a moment on this planet Earth. Who can envision my purpose from the beginning to the end? What shall I do, being lost within myself? Mixed with the raindrops, my mind falls to the Earth, only to be seen when the lightning strikes. What I shall be and what I shall find depends upon where the sunlight is leading me. Let me disappear behind what lies behind the rainbow to reappear in the raindrops that will fall.

# Sometimes

Some days are good; others are bad. Some shouldn't have happened at all. Sometimes, my mind is sharp; other times, it's dull. Maybe a nap should befall. Sometimes, I chase the dog; the dog chases the cat. Maybe I should put the dog outside. Sometimes, I watch the news, and I see the same political lies each day. Better to turn the TV off. Sometimes, I have a good idea, or my mind goes blank. Better not to think at all. I know the world is spinning around, and so are the troubles that lie within. Maybe I should hitchhike to nowhere. What are you expecting me to say that's significant? That's all.

# Society's Needs

The object of my affection is to reach a state of completion to advance society. Society is dependent on free thinking to advance science and philosophy. As ideas bounce from one mind to another, they become inflated to a new ideal. Involvement in the evolution of greater thinking will reveal many great things to be known. Thinking by ourselves in depth, going where nobody has gone before, going beyond the sunshine and the stars. Discussing and creating, discussing and creating—the greatest way to go. You may be right, or they may be right—the place to know. When mankind can increase his intelligence by not letting one erudite thought in his continual advancement be lost, that's where our goals should lead—to places never seen before.

# The Scientific Way

Don't start what you can't finish, for it will leave you in a state of limbo. A hypothesis about anything in reality can be plausible if supported by the facts supporting its inquiry. Thus, experimentation progresses until completed and, by mathematical proof, becomes significant and applied to reality. This is called the scientific method, for a theoretical concept to evolve in its application to reality. When I worked at the UCLA Brain Institute for three years, I experimented with this hypothesis. When one individual loses one of their sense modalities, the others will be increased in their capacity in relation to their environment. Proven to be significant mathematically, I stopped due to supporting a family. A great creative time of life.

# Thinking And Creating

Don't start what you can't finish, for it will leave you in a state of limbo. A hypothesis about anything in reality can be plausible if supported by the facts supporting its inquiry. Thus, experimentation progresses until completed and, by mathematical proof, becomes significant and applied to reality. This is called the scientific method, for a theoretical concept to evolve in its application to reality. When I worked at the UCLA Brain Institute for three years, I experimented with this hypothesis. When one individual loses one of their sense modalities, the others will be increased in their capacity in relation to their environment. Proven to be significant mathematically, I stopped due to supporting a family. A great creative time of life.

# Ridiculous Rhymes

Jack jumped over the candlestick; my middle name was Jack, and I never did such a crazy thing. Jack and Jill went up a hill to fetch a pail of water. What's so great about that? There are better things to know. Little Miss Moffett sat upon a tuffet. Anyone can do that; I'd rather play a game of checkers anyway. Humpty Dumpty fell off a wall and never could be put together again. Who knew that he was an egg after all? Eggs are usually eaten rather than falling off a wall; people don't care about putting them back together again. Mother Goose had so many children that she didn't know what to do; people with so many children go on welfare, and that's what they do.

# Our Crimes Living Things

All life must prevail and not suffer extinction by man. The roaring of the lion, the king of the jungle, and the chatter of the monkey where it stands. To enjoy the setting sun and the moonlight in the night. To hear the waves falling upon the shore and to see the bird sitting in sight. In seeing the great sea and the high mountaintops in view, maybe both are only viewed by just a few. Wake up from your dreams, ignoring all the cruelty created by others every day. Let no one escape from crime against other living things. Make them pay and pay every day.

# All Is Consumed

Most inanimate things rust or bust or disappear into the dust. Those rarely found are made of rare metals or hardwood. In a lost tomb or from beneath the ground, ancient artifacts in their entirety have been found. These findings are rarities, not commonalities, and are displayed in museums around the world. My major observation is that commodities are consumed and end up in some kind of yard. All are consumed by time, ending in a graveyard or as dust in the ground. To serve and be served in a positive way to live a prosperous life in totality. Life's only meaning is to prepare with love in knowing Jesus Christ as your savior and resurrection to eternal life. Eventually, all living and nonliving things shall be dusted in the ground, never to be found.

# Unknown Train

A whistle in the dark from some unknown train going to nowhere that I know. A freight or Amtrak train, I don't know, going fast to wherever its destination is true. I wanted to be a train engineer when I was young, going to places everywhere, blowing my whistle for all to get out of the way. Wearing your engineer's hat and pants, you shall take the train wherever it has to go. There is a crossing; I must blow the whistle that I am coming through. Track upon track, rumbling in the dark, a whistle I shall forget from a train I can't see.

# Existence Gives Meaning

We are infinity; our complexity gives our purpose to matter and energy to have been created at all. Our complexity in describing its existence, such as matter and energy in the creation of a star and galaxies, gives reason for it to exist. Man creates the math to substantiate it from its beginnings to now. While our universe has been expanding for possibly ninety-four billion years, going places unknown, man shall try to give theories for it all. Higher life forms give reason to an unending universe to befall.

# Ring Around The Posey

Ring around a rosy, a pocket full of posey, ashes to ashes, and we all fall down. A song by children during the Great Plague of the 1600s in London and the rest of Europe. What people didn't know at that time was that the fleas were carried by rats, as people were in close quarters bitten to cause the Black Plague. The rat is a carrier of the flea that bites its host, carrying the contagion that causes the deadly plague. Rosey, in the children's song, was the red rash around the black wound. Posies were the flowers whose sweet scent hid the stench of death. Ashes to ashes represented the cremation of the dead, and we all fall down, the mortality of all. Ring around a rosy, a pocket full of posey, and ashes to ashes, and we all fall down. Over one-half of Europe's population died of the Black Plague.

# Love

A faltering world must seek love to lift it up once again. It's love that shall cause a caring smile and the joy of living within. It's love that shall defeat hate, and people shall walk hand in hand once more. It's caring that shall open our neighbor's door with smiles that shall endure. It's kind words that will create good friends rather than enemies to enjoy each other in sound relationships. It's goodness and goodness that shall yield peace in a better world. It's love and only love that shall bring the sunshine to save a dark world, and people shall dance without fear. My heart yearns for love to abound everywhere, from the blue sky to the green grass below.

# When I Was Young

When I was young, I could see the beauty of a summer night that made my heart sing as stars appeared in the evening sky. Oh! Just bring back the full moon to shine upon my face once again. Just to hear a voice singing coming from the trees, telling how great life can be when you are young. I don't know what I shall do to bring back the blossoms' sweetness that fills the air everywhere. What shall I do in bringing your love back to my void heart and seeing your sweet smile again? Let me walk those paths of youth once again, that my heart will find your beauty once again.

# There Is Nothing

There is nothing one can do in hearing an ignorant world. Most of the events completed in a short life don't inspire a sorrowful world. Besides killing wildlife and destroying the trees and grass, throwing garbage into the seas. Contributing to extreme weather conditions breaking to new highs and lows. Air conditions are becoming worse and worse, and pollution of the blue skies is becoming grayer with our smoke. Trucks, cars, and our industry contribute to the smog that may contribute to our downfall. Does mankind have the intelligence to change anything? Possibly, everyone will need oxygen masks to breathe. Another beer and a tattoo can describe this generation best of all, leading to nowhere!

# The End

Sixty to seventy million people died in World War II because of the tyrannies of Hitler and Tojo. Now, the tyrannies of Russia, China, and more are creating World War III with much greater weapons to employ. All our problems of war and more lie in the hands of a world that does not understand! That's all and the possible end.

# Is There Justice In Killing

Is justice but a word that can't prevail over countries that have lost their freedom? Captured hostages are killed or treated with cruelty while being kept in a filthy prison. Wars are currently being waged with atrocities on both sides, with innocent citizens being killed with no regard for human life. Who shall disguise the dead bodies at the great cost of a state of war? This is humanity in the power and greed underlying it all. No matter what we achieve from our intelligent perspectives to create peace, it's basically war that we choose as a solution, after all.

# A Brand New Day

A brand new day comes, and then it goes beneath the moon before you can yawn. Mental time can be slower or faster than actual time, depending on your perception of reality. Let's say you're enjoying a hobby like building a model airplane; mental time will be faster than actual time. Let's say you're waiting at the dentist's office; mental time will be slower than actual time. Let's say you are on vacation in Las Vegas; mental time will travel faster than actual time. Let's say you are working hard on your job; the time never seems to end. The worst is if you get drunk and pass out; mental time can't be remembered at all. You won the jackpot; I shall count the money in actual time, gamble in mental time in winning or losing. Time to go to sleep, that's all.

# Under The Apple Tree

I sat below an apple tree and, in a distant sound, heard this old song. Don't sit under the apple tree with anyone else but me, with anyone but me. No! No! Don't sit under the apple tree with anyone else but me until I come marching home. I was very, very young when I first heard that great old song. It was a song written during World War II, talking about a soldier overseas fighting the war. In the song, he tells his girlfriend he left home not to sit under the apple tree with anyone else until he comes back to the United States after the war is won. Unfortunately, many soldiers never came back home, being killed in action. Those girls waiting just cried forever as their heroes lay buried in a grave far from home. But they could swear they still hear their voices singing as they sat under the apple tree, singing. Echoing in the distance, don't sit with anyone else but me until I come marching home!

# School Games

Remember in school when someone hid an object in the room, a person was chosen to find its hiding place? The farther you moved away from it, the classmates shouted, "Cold, cold." The closer you approached the object, the classmates yelled, "You are getting warm." When the object was found, the game was over. That happens in many incidents of real life; the closer you learn, the better you'll achieve. The farther you are away from learning, the greater the possibility of failure. Seek, and you shall find the object called joy and happiness in becoming an adult. If you know the difference between what is right and wrong, you shall create a good life. Be content and fulfilled in completing a short life.

# Creating A Good Life

What does it take to create a good life? Good health most of all. Success in creating a good return on money, living comfortably, and maintaining mental and emotional stability is important, after all. The ability to take a break with entertainment is necessary; it is a hobby to enjoy in spare moments to relax from the trials of it all. Consistency is very important in planning what is defined as a good life for everyone. Eating healthy foods and exercising are important. It's important to consult an expert when needed to help with problems you can't completely solve. A lot to properly organize in living a good life, so sit right down and start right now. Maybe I should take a nap and forget it for a little while.

## Our Country

Over here, over there, we shall come out of despair, and our country shall reign greater than ever before. All should fight for the right to show other countries we won't be pushed around anymore. Our country will demonstrate what freedom is all about and freedom of speech for our people to display. Don't let anyone retreat in portraying what hatred and tyranny are all about. All must cry out for liberty for all, including the whole world. The many promises that are made to create a greater America must be enacted and kept by those gaining power. One lie is enough to destroy a total agenda of truth. Let us rise up to protect our bill of rights against anyone who shall challenge them at all. Countries displaying nuclear weapons must be challenged without fear, in our wisdom, to overcome anything everywhere. Let the world know that our country is one of compassion for anyone seeking peace, but we must stand strong in what we believe. Most will attack us when they think that they will win. Let our military might be the greatest in the world. May our country return to believing in Almighty God and His truth.

## Not Knowing At All

Make haste in what you must do, even if you don't know what it is at all. If you get to where you are planning and don't know where that must be, be happy anyway. If you win first prize and don't know what it is, accept it anyway. If you complete something great and don't know what it is, celebrate anyway. If life has been good and you don't know why, give a smile anyway. If you have failed and cannot explain why, just don't look into the mirror anymore. If they say you're great and haven't found out why, just clap for yourself anyway. If people say you're on top and you haven't reached the bottom, just stand still anyhow. If people say anything at all, just pretend it's not true. That's all, that's all.

# Serving The People

In running to become a member of the House of Representatives from your state or a senator in the state you come from, being a member of Congress is a large responsibility in serving the people. This is what constitutes a representative government. Be sure you're serving the people and not yourself. There are state governments, including a governor who heads that state and senators and representatives who run the state as well. Dishonesty befalls when these elected officials put the lust for power and money over the people. When the supporters of honesty can't weed out the bad, our democracy shall fall. Justice in protecting our constitution for freedom for all, and our government is for the people, and by the people, it is best to endure.

# Failure: School And Colleges

About 5.5 percent of potential adults don't graduate high school in the USA. Most failures occur in Hispanic and Afro-American minorities, according to calculations, not with prejudiced intent. You say that's a lie instead of fact; that's up to you, but it's better to help anyone who's having trouble graduating high school. Of those accepted to a university, forty percent will drop out in the first year. In graduating high school, teachers and some parents help obtain a degree. In college, students must learn by themselves and study what they will need to learn. Most college students don't even know what to study for a given degree to achieve success. Many college entrants don't know how to study or are not prepared at all. Colleges want the money for the course; no mercy in how to pass them is given. Courses cost a lot, not including the books and living expenses, which are extremely high. Not being prepared and the costs are the reasons for forty percent of the failure. High schools must help the students wanting a college education to be prepared, and parents must start to save for the money needed, which is the highest in history. Maybe being prepared will stop fifty percent of dropouts from occurring, with only thirty percent failure. For America to be great again, education must endure and endure!!!

## Stopping Fear

The only thing to fear is fear itself, but fear cannot be stopped by words alone. One must consider what creates fear to abate what causes it for it to befall. Fear is one of the greatest emotions to be reduced for a better life; fear of failure is a good example. Education and good training to succeed reduce the fear of failure. Fear of heights and snakes is called phobia, which may require a psychologist to help. The fear with anxiety that causes emotional outbursts also requires help. Fear is something good in not getting hurt, like staying away from dangerous objects that can injure or take a life. Fear of doing the wrong thing instead of what is right, fear of facing hate instead of love. If fear causes you to lose sleep or hide in the basement, you need help. Fear or not to fear is the question of maintaining mental health. There are reasons to fear and reasons not to fear; you decide for yourself.

# Hey Bro

Hey bro, how are you doing? OK, I guess it is not working right now. Hey bro, how many kids do you have? Four, and working on five. I must call medical for more money and food stamps right now. Hey bro, are you working yet? Why work when the government shall pay for everything I need? Hey bro, want to meet for a hotdog or two and talk about nothing at all? See you there, bro, in my busy day of doing nothing at all.

# The Faster, The Better

Class and dignity must be taught for them to be performed, for they're way above the natural behavior that usually comes to be. Manners and respect come from people with high expectations who teach what is right from wrong. Children respecting their teachers and adults will mature in a successful direction, completing a better life. Learning how to think and presenting new ideas in communicating with each other. Reasoning and debating what is right or wrong. Education shall spread faster than the mind can follow in a better and safer world, too. Debates about how life began and where it might end, with the erudite winning the stage of life after all. The faster, the better, the faster, the better, for knowledge to become supreme, only beholden to Almighty God!!

# Political Conventions

Those two conventions, both Republicans and Democrats, have already chosen their candidates to try to become president of the United States. One candidate wants to make America great again; the other wants to enhance what he did before. The delegates represent each of the fifty states, yelling as chosen speakers reveal what their protocol for running our country should be. Borders, borders, and inflation are some of the issues of today. No one knows how much liquor has been consumed as they yell and jump up and down. Finally, the chosen vice president expounds upon the important issues to save our country. The chosen candidate to run for president expounds on all that is wrong with this country in total. He promises to fix everything for a greater America. The drunk delegates yell and clap at every word that is said and then go back to the states they represent. That's all.

# Going To A Restaurant With My Wife

I went to a new restaurant with my wife a month ago. I ordered a steak dinner with a baked potato sour, cream, and chives. My wife ordered a crab dinner, which suited her personality, with a large salad first. The food ordered wasn't as good as in the old days due to inflation and not hiring a good cook. I had to ask for a few toothpicks to get the meat out of my teeth. My wife, after eating the large salad, asked to take the crab home. I asked for a dessert menu, and the waiter responded, "No dessert today, or maybe a cup of applesauce." I inquired about the reason for no dessert, and he responded that due to inflation, they couldn't afford to hire a bakery cook. I received the bill, which was higher than before. The waiter said, "You can add a tip over here." I responded, "Can't add a tip due to inflation," as we left him with an angry face. Inflation and inflation, it's better to ask for a raise at work!!

# Take Ten

Take ten in taking a short break from the words that never stop moving my hand. Take ten in tasting a glass of expensive wine. Take ten in discussing trivial things that don't mean much in their simplicity. Take ten in a moment of sleep, lasting too short, falling off my chair. Take ten in having a short dream I can't remember at all. Take ten, take ten, in forgetting I took it at all.

# Nature Can't Do Anything

How could nature create elephants with trunks, giraffes with long necks, and zebras with stripes in such a unique way? Can nature create endless oceans, large mountain chains, and large green forests? Amazing to see. Can nature create the countless stars seen above and on the moon? Can nature create the sunlight and the rainfall that falls to the ground? Can nature create the flower that blooms and the trees for one to see? What can nature create? Nothing at all. Everything is created by the word of God!!

# Pompeii 79 AD

Pompeii, whatever happened to you? Was it a violent explosion of a volcano or an act of God that destroyed your people? It was an ancient Las Vegas of Rome, where the people of renown and wealth enjoyed every pleasure known to man. Bodies are preserved by the lava, with very few human remains lying within. This devastating event of the volcano devastating Pompeii occurred in 79 AD with little warning as ash buried the city. The people enjoyed great food and excellent entertainment, as well as prostitution. The people enjoyed bathing in the hot waters and the saunas that were available. Special plays were performed by actors, musicians, and dancers two thousand years ago. Proof of gambling was found in the city's excavation, and extravagant pictures were painted in many found rooms. The volcano erupted so fast that many couldn't escape death. Many new secrets are being found as the excavations continue to this day, revealing how the Romans lived two thousand years ago.

# Enjoy Life

And now I wonder what shall befall, what I shall say, in the lonely nights that lie before me. I don't recall all the moments that have passed away, in the great things that have happened to me. The blessings that brightened my life and the dreams that came true. There are some regrets, but many happy times cause a smile. Many good friends have passed away, but some still remain to enlighten my days after all. My wife is here to bring joy and happiness that are still to be found. The birds land upon my windowsill, and the sunshine appears, showing how beautiful life can be. Don't worry about a thing; I shall enjoy the rest that will come. Happiness can only be found in your heart, which receives it forevermore.

## To Have Or Not Have

To have not or to have, to have a little or to have a lot, to have much energy or to have a little feeling weak, to have to have everything under the sun. I hope that's all: to agree not to agree, to agree in peace or agree to have war, to agree to be friends, or to agree to be enemies. To become or not to become, to become stronger or weaker, to become lost or found, becoming dizzy in turning around. To win or lose, to lose or win, or do both at a given time, maybe I should sit down and rest. To shout or to yell, or to yell or shout, are they about the same thing? To advance or retreat, or retreat or advance, who has the better chance? To do this or that, I have had enough of those suppositions. That's all.

# The One I Adore

Only those sweet dreams can return the image of the one I adore, that beautiful face I can't see anymore. Going to the park and riding the roller coaster, screaming together as it goes up and down, eating an ice cream cone, and hearing laughter, how can those times be found? Laughter, talking, going to the movies, eating candy and popcorn—where has she gone? A void in my heart never to be filled. Audrey, just appear once more, once more, that my longing shall be stilled.

# Queen Of Football

In times that are gone, the queen of a football game was the most beautiful of them all. Marching with that beautiful face and great figure, twirling her baton in front of the great crowd. Crowned by the captain of the football team—what can get better than that for the one with the most beautiful face? At the present time, the queen can have just an average face and weigh two hundred pounds, jumping and twirling her baton. The captain of the football team was absent when she was to receive her crown. At the queen's dinner, she ate three plates of food. Things have certainly changed over time, but for me, I'd find a time machine to find the beautiful queen from a long time ago. That's all!

# Comedy Of Yesteryear

In the comedy of yesteryear, things that were funny are not seen anymore. Like hitting someone with a hammer over their heads when they do something dumb. Like pulling one's nose with pliers when they can't accomplish a simple thing. Like three people sleeping on a bed as they all fall to the floor. When trying to paint a simple wall, their brushes paint each other's faces by supposed mistake. One is the leader who beats up the stupid one for making the same mistakes. Like the stupid one trying to fix a shower leak, getting trapped within all the pipes not needed in completing a simple job. Smacks on the face and pulling hair in trying to get laughs in the comedy of yesteryear.

## A Four-Leafed Clover

I am looking for a four-leafed clover that is rare to be found. The first leaf is love, the second leaf is happiness, and the third leaf is the joy of living. The fourth leaf above all is you, whom I adore most of all. Every morning in my garden, I try to find the four-leafed clover that evades my searching eye. Suppose everyone would find that elusive four-leafed clover; what a great world this would be. I water the grass daily, hoping the four-leafed clover would suddenly appear from the three easily seen. So, I am still looking for my four-leafed clover for all my dreams to come true.

# Bitter Versus Sweet

Food that is bitter in taste is disliked and avoided by most, while food that tastes sweet is liked by most and purchased in great quantities. Likewise, a bitter person whose words criticize and attack others is usually left alone, talking to themselves. A sweet person who smiles is kind, compliments others, and is surrounded by friends. A person with a bitter viewpoint criticizes the government, accomplished works, and anything good at all. A sweet person applauds those giving to charities, creating artistic works, caring for the disabled, and making this world a better place after all. A bitter person walks with a frown, yelling in criticism of others, and is glad when people are sad. A sweet person walks with a smile, giving a kind word to others and sharing the joys of life with others. Bitter and sweet, bitter and sweet—which group do you belong to?

# Justice Or Injustice

Don't become alarmed when someone is pointing a finger at you when four fingers are pointed at him. You can be perfectly innocent in what is claimed against you when the one pointing is guilty without a doubt. What's this all about? Injustice can be disguised as the truth in this misleading world. Many have been hanged because of prejudice and hate instead of guilt in a world spinning in mistrust. Many don't care about justice at all, seeking revenge with hateful hearts. Jesus was crucified falsely in His perfect, loving life as Barabbas, the criminal, was set free. The majority of people screamed, "Free Barabbas, free Barabbas," as Jesus was condemned to crucifixion falsely for the forgiveness of our sins. Great people have been put to death, defending liberty and justice for all. Justice or injustice—only the brave shall stand for what is right in a world of lies. Let justice prevail over injustice to uphold the truth after all.

# Evil Knocking At Your Door

The blackbirds fly above with demons controlling their bodies, displaying the evil that abounds everywhere. Many black birds fly over the homes of people practicing black magic and witchcraft. Witchcraft is practiced by many, who use curses to harm people they hate, creating pain or even death. Using special chants from their evil books, they cause enemies to fall. Chanting for demons of hell to possess anyone they choose, which may be you. Voodoo is practiced in powerful dances to please Satan himself, receiving his evil power. Curses with chicken bones at your door mean your death shall befall. Those who use tarot cards to read their future and some who use crystal balls try to see what will happen to them. Many playing these dangerous games don't know the evil they perform. It says in the Bible that there are evil spirits that dwell in high places. People become aware of the evil that abounds everywhere, even in their beds or next door. Watch out, watch out, when you hear the dog howling in the night.

# The Spirits Scream

In the cold of the nightlies, the tomb in an old cemetery was forgotten by the living long ago. In the tomb lies the restless spirit in the anguish of once living a very sinful life. No repentance for the horrible murders once performed, no conscience in shedding blood. Every night, you can hear him scream in agony for his lost soul. In the houses nearby, those who hear those horrible screams hide in their beds in fear of being snatched into hell. These spirits scream beneath the moon in their graves until the rise of the sun. Their lifeless forms are seen walking slowly in the lost cemetery. Just one moment of life to ease the agony of eternal strife, never to see the light anymore. Many buried in this cemetery committed bloody murders with evil beyond repentance. The living hear the screams of the dead louder and louder in the horror of it all. Stay away from the one in the tomb lest he snatch you into his grave forevermore. Beware, you don't repeat the evil they did, and your soul joins the screams of agony beneath the moon, never to stop, never to stop. Maybe you shall hear them tonight.

# Hard Chores To Do

Wash the dishes, clean the floor, and throw the garbage outside in the cans. I had to hurry; I had to go to work in an hour or two and change my clothes. Maybe I should marry to help me with everything I must do in this empty old house. Even the simplest things become a chore, and when the hardest shall befall, what shall I do? Maybe hiring someone to help is cheaper than a wife; I haven't made up my mind. I must reach beyond the garbage to greater insights before I go out of my mind. I must think about going beyond my door to work, which I don't like. What a life, what a life, as I beat my head against the wall.

# A World Of Dismay

What is the solution to the problems of this world that nobody can solve? Not even a large country can stop a war or make a poor country become rich. Not stopping a country living in tyranny from becoming free or a country living in starvation from having food. Who can give a reason for bondage and hatred to be chosen over liberty and freedom? Don't know. Can anyone explain why murder is allowed in some countries with no investigation to befall? How can you explain why everyone doesn't stand against the evil that does befall? Is everybody hiding in their bedrooms, not fighting for what is right, afraid of their shadows in the darkness of the night? Don't know anything else to say in a world of dismay, just to pray and pray.

## Where To Find Money

Money cannot be found in a money tree in your backyard. Money can be found in hard work or in sound investments in the financial market. Money can be found in opening a new business or investing in real estate. Don't count on anybody giving you the money to do anything. A wise investor should get the advice of an expert and double-check what they have to say. Money is like paper in your hand—when dropped, it can be blown away. So, if money makes the world go around, let it spin in the direction of me. Bill collectors say, "I want my money, I want my money," screaming in your ear. Be happy that you are living through it all. Isn't life more important than money? I DON'T KNOW, I DON'T KNOW!!

# A Penny For Your Thoughts

It was once said, “A penny for your thoughts,” but now it has become a dollar with the inflation that has occurred. It is said, “To be or not to be,” and I don’t care while drinking my cup of coffee. It is said, “That music shall quiet an anxious mind.” I don’t care if it’s music I don’t like. It is said, “If you conquer, you must divide.” I do care when you give me the best side. It is said, “Take pride in the work that you do,” but only if I can take a nap if it turns out bad. It is said, “That two is better than one,” but it depends on how good the other one can be. It is said, it is said—who cares at all?

# Whistle While You Work

Whistle while you work, thinking you need to buy a new shirt and a pair of socks. Wait! After paying all the bills, I only have enough to pay for one. That's all. Either get another job or sew the holes in the socks that I have on. My credit cards are at their limits, and I can't make any loans—maybe sell my soul for whatever it's worth? Inflation has dug a deep hole that I can't climb out of. Will someone bring me a glass of water? My dreams are shattered by quitting school and not becoming a licensed painter or plumber. Being married with kids that will come doesn't look like a future possibility. Marrying a rich woman isn't possible for one who can give little in return, making twenty dollars an hour and no future at all. Maybe, with luck, something unknown shall deliver me from this lowly state? Dreams of success that shall never be seen in reality after all. I feel that I am locked in a cage without a key to escape. That is all that there is to say and say in living one's mind.

# New Thoughts That Abound

Searching in schools for those who are excelling in higher learning in science and humanity. Without their success, the majority would go around and around in a circle of ignorance. The complexities of reality shall be revealed only by astute minds, giving meaning to what hasn't been found. New theories must replace the old ones being dispelled, with the erudite minds seeking the facts hidden before their eyes. Searching and searching with profound thoughts, creating a whole new world never before seen by anyone. The majority sit in apathy, content in looking at the ground. Scientists and humanitarians go forward in creating a better world for everyone with new thoughts that abound and abound.

# Mixed-Up

You can be wrong when you think you are right and dumb when you are bright. Mixed-up, so am I. The wise shall become ignorant, and the powerful become weak. Still mixed up, so am I. The drunk say they are sober, and the sober say they don't drink. Still mixed-up. The rich become poor, and the poor remain poor. How do you know? The innocent become guilty, and the guilty are released from jail. That is only partly true. Some failed the test, and others had to take it over again. Need to explain more about that one. Some recite poetry they think they comprehend, while others just recite words. Maybe both should be reading comic books. People laugh at a joke they don't understand, while others don't know it is a joke. Maybe the joke was on me. That's all, folks!

# Consumed Things

Time keeps marching on, creating memories from those moments that just occurred. My brand-new car is standing in the junkyard, and my new suit doesn't fit anymore. What else can time take away from me? My pants don't fit what I bought five years ago, and I can't run as fast as before. My wife can't chew gum because her false teeth will fall out. Time, you have no guilt in the things new becoming old. My feet hurt from walking too long, and I slept ten hours instead of eight. I can't find things sitting before my eyes and think it's night when it's daylight. Time, you just keep marching along. I think it's dinnertime when it's breakfast after sleeping too long. I was fired from my job for coming late when I thought I came on time. Time keeps marching along—I better enjoy everything that I can as the hours keep flying away. The time you keep ticking away without return. Goodbye and goodbye!

# Junk Food

McDonald's, Jack in the Box, and Burger King are fast food places that will reduce your life span by twenty percent. A double cheeseburger with bacon, pickles, and the rest of the toppings, with French fries, will make your stomach want to run away. Next is the hot Mexican food, tacos, and burritos topped with hot sauce—take another five percent from your life and visit a doctor for gas and acidity. Popeyes and KFC with fried chicken contain large amounts of fat and sodium to give one a heart attack. Chinese food with large amounts of fat and salt is best sent back to China. A hot dog or two with mustard and relish, with a Coke, is enough to get anyone sick and shorten your life again. The great-tasting pizza with calories and sodium and the great toppings that a dog or cat should eat. See you at the emergency clinic after eating fast food. Thanks, and thanks for not drinking that healthy green stuff that tastes so bad.

# Love Calling To Each Other

The love call that falls upon the breeze and is carried by the moonbeams to the one meant to hear. Not even death shall quiet the love call that is heard beyond the mountaintops and above the clouds, called to a lost lover anywhere to be found. The love between two that's never-ending shall reach eternity and beyond. Lovers shall be calling each other in a time of need to change a tear into a great moment of joy. Lovers calling each other with voices echoed into the air for all the lovely words to be heard everywhere. I shall be calling you, and you shall be calling me in voices forever heard.

## Beautiful Lady

Weep not, my lady, for his love did fade away.

His smile and cares, gone and didn't stay,

Tears shall fade, and new sunrises will bring,

New romances, a suitor's hand, and of unique style.

Beauty shall be your partner in just a few years,

Display it properly to bring happiness and erase the tears,

Romance springs forth, found only on that windy hill,

Meeting each time in that hidden place, the greatest thrill.

Romeo met his beloved Juliet, a story forever being told,

The folly of misadventure, of tragedy forever to unfold,

In the intensity of love's wonders there,

And no words to be, these stories of love shall not be retold through history.

My lady, beware the evil in men,

Putting away romance when his loving tune is over.

There are second chances, your tears shall return, my lady,

When love is not there, it will win.

The cares of love, nothing there to win.

# A Poem

A poem is leaving your door, adventuring to see more. Traveling roads, dining with one you adore.

A poem paints a picture, carried within, Among people and places, diving deep down, Swimming oceans under the sun, matching birds in sky's high places.

A poem reflects, greater comprehension, an art not to abate. Sky deeper blue, sun brighter ablaze, creating worlds anew.

A poem flows, an unending river, fish swimming below, Embellishing history, enriching today, forging a brighter future's to glow.

In reality, a poem may fantasize or glamorize the world where we reside.

Flashing insight, thunderous judgment to decide.

A poem is a youthful gift, enlightening truth.

Seeking blessings in deeds, comforting feelings described as mere.

A poem stirs emotions, in simplicity or complexity.

Profound in its depth, simple as the wind, gentle as raindrops falling upon in a crowd.

Poem's eternal soundness, seen in rainbow's beauty.

Goodness of love, universally agreed to be known.

# The Chance For The High Finding The Ground

Find two snowflakes in a pattern. Are they the same?

Promise peace, never war, nuclear weapons ashore,

A dream coming true, not lost in the night.

Peace, justice, and liberty for all but me.

Fantasy in part becomes reality in time.

Kings and queens exist only in a deck of cards,

Troubles are for some, others have none.

Without a sail, a boat turns around and around.

A song to sing, a bell to ring.

What does that mean?

We are what we are not and become what is not.

More, not less, will state the rest,

The end will come only to begin again.

# Spring Time

A time of colored bushes, a time for green.

A time of showers and the birds to sing,

A time of birth and new habitats for life to grow.

A time for the creeks to run and rivers to flow.

A time for the ant hill and the bees and their honey,

A time for the rainbow and the hopping bunnies.

A time of new life appears under the brightness of the sun,

A time that the wolf goes home and the wolf does howl.

A time of romance holding hands in the month of May.

A time when springtime passes fast, and summer comes our day.

# Never A Generation

Never use words that are profane or cause our fuse to burn, For profanity expressed in tongue, defamation is never right. Never use tears to be at war's end in summation of our flight, By rationalization on making a failure be a greater insight.

Never make a covenant that can be broken and can't be kept, For those who break it will find no repentance in a time of need. Never delight and boast about treasures lying behind your door, Coveted by neighbors even more.

Never eat of the tree of good and evil inhabited by the evil one, For the penalty in it can be greater than the gun. Never want to hear the screams of the souls of Satan cast into hell, But put on the Holy Spirit's armor Satan cannot overpower at all.

# Joy

An expression of pleasure in a task well done, Undaunted behavior in happiness as a group becomes one. Joy to the world, even if it will become a moment to last, With people holding hands rather than fighting is everyone's task.

Differences are not judged merely from face-to-face. Joy in all coming together to advance the human race. Goodness and evil reside in all people from below and above. Salvation in Jesus and forgiveness in his shed blood. Be prepared to speak good words to all that you shall meet, Caring for others' needs touches all that you shall greet.

Let gladness overcome sadness in events both warm and cold, Meeting the challenges with fairness in a heart of God.

# School Days

Gone are the school classmates, only as apparitions in my mind, Searching for those faces that I shall never again find. Neither school nor student can be found under the shining sun, Someone must lead me to a place where is even one.

Voices long to be heard and faces seen, disappearing with time, Nothing is left of my books and pencils that were once mine. Nobody now reflects upon school days they did learn, Upon the various classes taken and the grades they did earn.

What joy in hearing their youthful voices calling my name, Greater is their friendship than all the riches or hall of fame. The bell has rung, classes to change, waiting for a pretty girl to take her seat, No greater play in a scene to complete, memories of those days to repeat

Good teachers of geometry and English no more seen to teach,Exemplified teachings illuminated my mind, beyond reach. Only pictures of classrooms and students, and a last word, All their voices testing their existence are gone, never heard again.

Made in the USA
Monee, IL
10 November 2026

7842b509-0155-481c-9d2a-d4b5602c9f0cR01